The Deliberative Practitioner

The Deliberative Practitioner
Encouraging Participatory Planning Processes

John Forester

The MIT Press
Cambridge, Massachusetts
London, England

Third printing, 2001

This book was set in Sabon by Achorn Graphic Services, Inc. and was printed and bound in the United States of America.

Library of Congress Cataloging-in-Publication Data

Forester, John, 1948–
The deliberative practitioner : encouraging participatory planning processes / John Forester.
p. cm.
Includes bibliographical references and index.
ISBN 0-262-06207-0 (alk. paper). — ISBN 0-262-56122-0 (pbk. : alk. paper)
1. Regional planning—Citizen participation. 2. City planning—Citizen participation. 3. Political planning—Citizen participation. I. Title.
HT391.F65 1999
307.1′2—dc21 99-15445
CIP

To Shimon Neustein,
a friend of words and wonder

Contents

Preface

This book grew largely from essays that I wrote first for formal lectures to planning schools, workshops, and research conferences. They reflect less any holistic scheme or grand plan than a pragmatic, continuously improvised, intellectual development, with each essay's new difficulties helping to define the agenda for essays that followed. If I do not present any one grand scheme here, so much the better. But I do develop variations on several themes that integrate practical and theoretical issues of public participation, political deliberation and learning, and city planners', public managers', and public policy analysts' possible work. The abiding questions that motivate everything that follows ask how diverse community members and the planners hoping to work with them can act more effectively in the face of political inequality, racism, turf wars, and the systematic marginalization and exclusion of the poor.

In a sense, this book explores an odd set of questions: When planners are acting well, just what are they doing? How can the probing, learning, and innovation of extraordinary practice draw on the basic elements of quite ordinary work? This narrows the focus of the book a great deal, and it explains, for example, why this book is not about geographic distributions, architectural styles, or the history of urban development. This book is about people working with others to attempt to remake their common future. My concern throughout is with deliberative and participatory practices: inquiring and learning together in the face of difference and conflict, telling compelling stories and arguing together in negotiations, coming to see issues, relationships, and options in new ways, thus arguing *and* acting together. The subject is practical public action in messy political circumstances, and my objective is to illuminate both requirements and opportunities for productive, if inevitably political, deliberative practice.

In part, I have come to see that a great deal of this book explores problems of the micropolitics of practice that I first wrote about as the daily political work of critical listening, in a paper aptly subtitled "Critical Theory and Hermeneutics in Practice" (1980). I had been interested, coming to doctoral work, in the ways that planning analysts could be anything more than ideologues working for government. That interest led me to, and then through, the study of philosophies of science: Could objectivity simply be whatever one wanted it to be, as facile social constructivism seemed to suggest? Hardly, as Hilary Putnam shows us, more clearly than anyone else perhaps. But social constructivism and then postmodernism did have a deep point: social beings construct their meaningful worlds through language and myth, ideology and tradition, through systems intertwining knowledge and power. So I turned to the problem dividing phenomenologists and critical theorists: How to integrate a powerful understanding of hermeneutics, of the importance of interpretation, with an analysis of systematic relations of power that illuminated domination and oppression, resignation and hopelessness? Seeing that social action shaped and claimed meaning through practical performances of speech, writing, gesture, or ritual helped me to link hermeneutic and critical theories, for in practical, rhetorical action, social and political actors not only made "meaning," but enacted complex relations of power, for example, by selectively setting agendas. Selectively shaping attention seemed to be the basic work that linked the work of phenomenologists (who studied attention, meaning, and social constructions) and that of critical theorists (who studied selectivity, ideology, inclusion and exclusion, power, and representation). But this led to further problems of evaluation and ethics, selective narrative representation and storytelling, that Martha Nussbaum's striking work on literature and the moral imagination made clear: when planners spoke selectively, when they told these stories rather than those, they did complex and very practical moral and political work. And so this book takes off from there, probing simultaneously interpretive and practical, political and ethical problems of public deliberation and deliberative practice, being able to work effectively with others, others who are different, in many fields of professional work.

Because I work in Cornell University's Department of City and Regional Planning, the chapters that follow deal for the most part with public-serving urban planners and policy analysts concerned with urban design and housing, land uses and transportation, jobs and the environment. I must ask others in related, applied fields like education, social work, and public health to judge the ways that deliberative practitioners in their fields respond similarly to, or differently from, the practitioners in the United States and abroad whom I have profiled and discussed here.

The chapters draw from literatures in political science, law, philosophy, literature, and planning, predominantly, but for the most part they follow my earlier strategy of putting the theory talk in the notes. When a doctoral student at Cornell, Robert Letcher, paid me the high compliment of the criticism that *Planning in the Face of Power* was too practical, I thought that here was a problem I would not mind having. I cannot predict if this book will have the same problem, but I have hoped to address practice, in a range of fields, in a way that practitioners and theorists alike can appreciate. I have tried to show how insightful practice can lead to stronger and deeper theory, and I argue that students in applied fields need to consider more carefully both the insights of practitioners and the questions that cogent theory poses for us. Nevertheless, we need to be vigilant for the hype and rhetoric coming from theorists no more than practitioners, the rhetoric that ends up numbing rather than teaching us, intimidating rather than encouraging and enriching us. Whether we face the rhetoric of "empowerment" or the "free market" in the field or "discourse" and "hegemony" in the academy, we need to keep our eyes on actual practices struggling in the face of power and inequality to build better cities and healthier and more vibrant communities, not shoving distinctions of better and worse, or true and false, under the thin rug of new academic fashions. This means too that we cannot study practice without studying ethics (creating or squandering value, achieving or repressing autonomy, and so on) at the very same time—and so the chapters that follow work to integrate theory and practice, pragmatism and ethics as well.

Not long ago, I had the great good fortune to visit Dino Borri and Angela Barbanente in Bari, Italy. Dino was completing his Italian translation of *Planning in the Face of Power*. Not only were my hosts gracious and generous beyond words, but the questions of their students were a further gift. Working with a small and very ancient town to the south, Marika Puglisi, Adele Celino, and Grazia Concilio pressed questions about participatory planning, the possibilities of mediated negotiations, the viability of community-based, participatory action research. How could they approach their problems, they wondered: to work with current residents, to leverage monies from the government, to preserve historic buildings, to promote tourism—all without endangering the townspeople's ways of life? As we discussed these questions, I first realized how the essays I have reworked here could fit together into a coherent whole, and a practically pitched whole at that. So I am very happily indebted to Dino and Angela, and Marika, Adele, and Grazia too, for provocative conversations that led me to restructure this book.

Each of the chapters that follows has been revised from previous incarnations as lectures, working papers, or earlier articles. Chapter 1 evolved from lectures at the University of Illinois, Urbana-Champaign, the State University of New York at Buffalo, the University of Puerto Rico, and Cleveland State University through a chapter in the book Frank Fischer and I edited, *The Argumentative Turn in Policy Analysis and Planning* (1993). Chapter 2 evolved from an earlier version in Mandelbaum, Mazza, and Burchell, eds., *Explorations in Planning Theory* (1995). Chapter 3 grew from lectures at the University of Nebraska, Lincoln and the University of Tennessee, Knoxville, a working paper of the Negev Center for Regional Development of Ben Gurion University, Beersheva, Israel, and a revised article in the *Journal of Architectural and Planning Research* (1998). Chapter 4 has been revised from "Democratic Deliberation and the Promise of Planning," the 1995 Lefrak Lecture, University of Maryland, College Park, delivered on November 8, 1995. Chapter 5 was written while I was Lady Davis Visiting Professor, Faculty of Architecture and Town Planning, at the Technion, Haifa, Israel, during 1993–1994. First presented at the Workshop on Evaluating the Theory-Practice and Urban-Rural Interplay in Planning, Bari, Italy, November 18–20, 1993, an earlier version appeared in Steven Esquith, ed., *Demo-*

cratic Dialogues: Theories and Practices (1996). A substantially abridged version appeared in Dino Borri et al., eds., *Evaluating Theory-Practice and Urban-Rural Interplay in Planning* (1997). Chapter 6 has been adapted from "Envisioning the Politics of Public Dispute Resolution" in A. Sarat and S. Sibley, eds., *Studies in Law, Politics, and Society* (1992). Chapter 7, originally prepared for the Workshop on Critical Theory, University of Minnesota, Duluth, October 17–18, 1994, and the Association of Collegiate Schools of Planning Conference, Phoenix, Arizona, November 5–8, 1994, is deeply indebted to the pioneering work of Dan Bar-On (1990, 1999). Chapter 8 evolved from a paper prepared for the Conference of the American Planning Association, Washington, D.C., May 9–13, 1992, and the Workshop on the Evaluation of Comprehensive Urban Planning, Umeå, Sweden, June 1–2, 1992.

This book grows out of varied sources of empirical material. The material from first-person accounts—the profiles of planners—comes from focused, taped interviews conducted by several graduate research assistants (Linda Chu, Erika Lund, Fred Rose, Kathy Seeburger, and John Welsh, in chapters 1, 2, 5, 7, 8), by myself and Raphaël Fischler (interviews with David Best and Arie Rahamimoff in chapters 3 and 4), or by myself alone (interviews with Rolf Jensen, chapters 3 and 4, and Baruch Hirschberg, chapter 4). The transcribed excerpts of planners' meetings and conversations come from several years of my tape recording and field observation in a small city's planning department (chapters 1 and 2). My account of Mary Jo Dudley's work in chapter 7 comes from my notes from a seminar she presented at Cornell, and she has checked that account. I am grateful for her time, insight, and assistance. For their generosity and accessibility, I am deeply grateful to planners, who remain anonymous, in the greater Boston metropolitan area, and especially to the staff of the City of Ithaca's Department of Planning and Development. For many years, Ithaca's planning staff has welcomed me as the "fly on the wall," even as they and a series of mayors have come and gone. Thanks to Matthys Van Cort, Paul Mazzarella, Herman Sieverding, Kathy Evans, Helen Jones, Glenn Goldwyn, Doug Foster, Leslie Chatterton, Tricia Norton, John Meigs, Peter Weed, Linda Tsang, Paul Stephenson, Marie Corina, Ev Hogben, and Eileen Jacobs.

I am grateful to many kind colleagues and students for their thoughts and suggestions regarding these earlier conference papers and lectures. Patsy Healey and Charles Hoch, in particular, built directly and critically on several of these chapters in their published work, and I am grateful for their continually thoughtful comments and suggestions. For their encouragement and critical comments on individual chapters, I am grateful to Greg Alexander, Ernest Alexander, Judith Allen, Rachelle Alterman, Dan Bar-On, Howell Baum, Bob Beauregard, Juliana Birkhoff, Dino Borri, Craig Calhoun, Penelope Canan, Rafael Corrada, Naomi Carmon, Stephen Chilton, Dennis Crow, Barbara Czarniawska-Joerges, David Deshler, David Godschalk, Kieran Donaghy, Yehezkel Dror, Andreas Faludi, Raphaël Fischler, Ann Forsyth, John Friedmann, John Gaventa, Silverio Gonzalez, Davydd Greenwood, Jennifer Greene, Tom Harper, Jean Hillier, Elizabeth Howe, Judith Innes, Rolf Jensen, David Johnson, Karen Jones, Alexis Kaminsky, Wendy Kellogg, Kristian Kindtler, Abdul Khakee, Deborah Kolb, Norman Krumholz, David Laws, Hubert Law-Yone, Bob Letcher, Morten Levin, Mark Lindemann, David Lyons, Seymour Mandelbaum, Ann Martin, Jim Mayo, Luigi Mazza, Beth Meer, John Nalbandian, Shimon Neustein, Bill Potapchuk, Arie Rahamimoff, Ken Reardon, Fred Rose, Shoukry Roweis, Tore Sager, Leonie Sandercock, Annette Sassi, Janet Scheff, Lynda Schneekloth, Donald Schön, Robert Shibley, Steve Shiffrin, Deborah Shmueli, Henry Shue, Susan Silbey, Stan Stein, Ernest Sternberg, Lawrence Susskind, Gilles Verpraet, Ariella Vrnesky, and Oren Yiftachel. In 1993–1994, I was lucky to be supported by a Lady Davis Visiting Professorship at the Faculty of Architecture and Town Planning, Technion, Israel, where the graciousness of my hosts was unsurpassable. For transcribing help in Israel, countless thanks are due to Elizabeth Falcão.

For many years I have had the good fortune to work with wonderful Cornell students, graduate and undergraduate alike, who have helped me with interviewing, transcribing, and, more recently, editing telephone interviews with planning practitioners. I am grateful for help from Linda Chu, Brian Kreiswirth, Kathrin Bolton, Kathy Seeburger, John Welsh, Jessica Pitt, Charles Loeb, and Erica Lund. For support of several years of such interviewing, I am in debt to the Cornell President's Fund for Innovations in Undergraduate Education. I am grateful too for support for our interviews with participatory action researchers, support that came from the Center for International Studies' project on Social Research for Social Change, less formally known as the Cornell Participatory Action Research Network. Thanks go as well to Cornell alumnus Mario Rothschild for his generous support of our efforts to collect practitioners' stories.

But more than anything else, the continued encouragement and healthy senses of skepticism and good humor of Betty, Kate, and Daniel have made this book possible.

Introduction: Renewing Planning Practice by Fostering Public Deliberations in an Adversarial World

Because planning is the guidance of future action, planning with others calls for astute deliberative practice: learning about others as well as about issues, learning about what we should do as well as about what we can do. So when city planners deliberate with city residents, they shape public learning as well as public action. Sharing or withholding information, encouraging or discouraging public participation, city planners can nurture public hope or deepen citizens' resignation. Working from the accounts of practitioners in urban and rural settings, North and South, this book shows how skillful deliberative practices can encourage and carry out practical and timely participatory planning processes. In so doing, this book takes planning practice as a bay window onto the wider world of democratic governance, participation, and practical decision making.

Working to promote jobs and housing, a healthy environment and better schools, city planners—and their first cousins, public policy analysts—have complex and fascinating jobs to do. They must study complex social, physical, and economic problems to propose effective strategies of public response. They must listen carefully to work through controversial political arguments. They must learn not only about "the facts" at hand but inquire about value too, asking what ought to be honored, protected, sustained, or developed—what, practically, should be done. Not least of all, they must inform, advise, and even coach a range of public officials and appointed, elected, and grass-roots decision makers too. These ordinary challenges of planning are actually quite extraordinary. They can teach us about the theory and practice of democratic politics, public management, and the public-serving professions more generally.

Providing more than an overview of the chapters to follow, this introduction characterizes the distinctive approach of the book, its bias for practice, and its motivating concerns (or more stuffily, its "normative bases"). Not least of all, this introduction suggests how this book does part of what it says, asking readers to interpret carefully and insightfully, just as planners must.

Deliberation and Learning in the Midst of Adversarial Political Actors

As planners work in between interdependent and conflicting parties in the face of inequalities of power and political voice, they have to be not only personally reflective but politically deliberative too. Just as "reflective practitioners" learn from experience (Schön 1983), "deliberative practitioners" work and learn with others. To show how a deliberative planning practice can be pragmatic and politically critical at the same time, this book interweaves planners' own stories and insights with commentaries that draw lessons from these actual cases.

This book takes a fresh approach to the study of planning practice. I do not contrast planning to markets, nor do I equate planning with either rational decision making or its more descriptively realistic "incrementalist" alternatives. Instead, I explore planning and policy analysis practice as it responds to the pervasive challenges of political-economic and social interdependence. Because so few political or economic actors can act unilaterally, all by themselves, planners and public policy analysts typically work in between these interdependent and often conflicting parties. The state wants transportation improvements; the neighborhood residents want less traffic and safer streets; environmentalists want to protect open space and easements; housing advocates want to encourage affordable housing; and so on. Because local politicians may interest themselves in many of these demands, though, planners have to work with many of these demanding parties at the same time—parties whose mutual distrust and strategic posturing regularly undermine their collaborative problem solving.

In cities and regions, neighborhoods and towns, planners typically have to shuttle back and forth between public agency staff and privately interested parties, between neighborhood and corporate representatives, between elected officials and civil service bureaucrats. They do not just shuttle back and forth though. Trying to listen carefully and argue persuasively, they do much more. They work to encourage practical public deliberation—public listening, learning and beginning to act on innovative agreements too—as they move project and policy proposals forward to viable implementation or decisive rejection (the "no-build" option).

Unlike many other professionals, planners and policy analysts have to be astute bridge builders, negotiators, and mediators at the same time (Forester 1989, Susskind and Ozawa 1984). As they try to see past problems and future opportunities through the eyes of many different actors, planning analysts try to build critically informed but pragmatically viable agreements. Working in between many affected parties or stakeholders, then, planners and policy analysts face a pressing and central challenge of democratic politics: the challenge of "making public deliberation work," making participatory planning a pragmatic reality rather than an empty ideal (Hoch 1994, Healey 1997, Gutmann and Thompson 1996, Benhabib 1992, Bessette 1994:46).

In chapter 1, we see immediately how practical deliberation in city governance involves astute listening, political reflection, and the appreciation that apparently innocuous storytelling on the job can do a great deal of work: framing agendas of discussion, shaping reputations, identifying important new issues, and more. Sitting in on a planning staff's conversation, we see how they evaluate and learn about the efficacy of their work. We hear too from a young city planner who is learning about the real politics of her job for the first time. We turn in chapter 2 to learn from an embattled community planner and a courageous African American housing counselor he hired to work with poor whites seeking home ownership. We learn from their story how planners must develop an astute practical judgment to deal with far more than "the facts" at hand, especially when they face economic and political uncertainty, cultural and gender differences, and racially charged legacies that threaten planning processes and outcomes. In these opening chapters, we see how planners can intertwine reflection and deliberation in politically realistic and effective practice. We see, too, how deliberative practice can integrate pragmatism and vision.

In chapter 3 we explore urban design and politics through the story of an Israeli planner-architect restoring the Old City and developing the modern city of Akko at the same time. We see how planners' spatial analyses inform the deliberation and participation of many parties, how planners may build a design consensus piece by piece through facilitated processes that blend architectural, cultural, political, and intensely bureaucratic concerns. In chapter 4 we learn from senior Norwegian and Israeli planners and architects as they negotiate over housing and the environment, over spatial and industrial-economic concerns all at once. We learn here too of the importance of "diplomatic recognition" (taking the other seriously) in the planning process, the work that such recognition requires, and the practical exploration, the deliberative crafting, of the policy and design options it makes possible.

In chapter 5 we complement the stories of community planners, participatory action researchers, in rural Venezuela, desperately poor East St. Louis, rural Norway, and a major U.S. city to argue that astute planning requires much more than the talk of dialogue and policy argument. Without supportive and safe settings that foster mutual recognition and respect, practical storytelling, surprise, and insightful listening, planning processes will degenerate into knee-jerk "us against them" adversarial bargaining. But when planners can facilitate processes of multiparty inquiry and learning, trust and relationship building, public participation can produce not just noise but well-crafted practical strategies that address real needs. Chapter 6 refines this view of participatory, deliberative processes by examining the politics and expanding practice of mediated negotiations, and activist mediation, in public and environmental dispute-resolution processes.

Throughout these chapters, we extend the individual work of reflection-in-action toward the more social and political work of practical deliberation. Reflecting alone, a practitioner learns; deliberating with others, practitioners learn together and craft strategies to act collaboratively. But how can parties do this when they distrust or even detest one another, when they inherit painful histories of eviction or terror, racial hatred or sexual violence? Chapter 7 turns to the work of planning activists in Colombia and in Boston—and to recent Holocaust historiography—to understand better how public deliberation requires the recognition and even working through of abiding loss and traumatic histories. In chapter 8 we return to the story of a local planner who is facing pervasive pressures of urban governance: corporate and public demands that create the day-to-day ethical challenges of city planning practices.

We learn in the accounts that follow about the possibilities, requirements, and risks of deliberative practice from city planners and urban designers, architects and "community development" consultants, community activists and activist-scholars. These many kinds of planners advise politicians, negotiate with private developers and public agencies, and try to respond to residents' hopes and fears. As we shall see, these planners can also manipulate poorly organized publics or encourage public voice and expand citizens' opportunities to improve the environment, the economy, and our communities (Lowry, Adler, and Milner 1997). In a phrase, planners can be hired guns or public educators (cf. Yiftachel 1995 with Hoch 1994; Reich 1988).

This book focuses on those planning practices that seek to expand practical democratic deliberations rather than to restrict them, to encourage diverse citizens' voices rather than to stifle them, to direct resources to basic needs rather than to narrow private gain. Although many of the following planners' stories will be plausible and realistic to many readers, they will nevertheless suggest skills and sensitivities not well cultivated in our schools. So these chapters seek to encourage better planning by showing what is possible in practice, not what is typical (cf. Teitz 1996).[1]

A Distinctive Approach

This book complements but also differs from the work of colleagues concerned with urban governance, policy analysis, and planning practice. For the sake of contrast—but at the risk of exaggerating differences from rather than assessing similarities—with authors with whom I share a great deal, I will suggest what may be distinctive in the chapters that follow. Where some authors focus on psychological aspects of planning, I focus on planners' more social and political interactions (cf. Baum 1990, Moore 1995). Where some focus on meaning, I focus on the deliberate or even strategic production of meaning, along with its daily interpretation in the political and ethical work of listening critically and developing practical judgment (cf. Marris 1996, Yankelovich 1991). Where some focus on power as structuring and limiting (Yiftachel 1995), I take power to be enabling as well, a politically shifting relationship rather than a fixed position or possession. Where some focus on ethics as a system of rules or codes, I take ethics to encompass the allocation and recognition of value, so I understand ethics not as standards to follow, but as pragmatic action always done well or poorly, always potentially assessable by standards, consequences, and qualities of action (virtues) (cf. Hendler 1995, Rorty 1988, Nussbaum 1990). Some take planning theory to be about intellectual history (Friedmann 1987) or decision making (Faludi 1996). I take it to assess planning as deliberative action that shapes others' understandings of their cities, their selves, and, crucially, their possibilities of action, for better or worse. As important, I take planning theory to finish the sentence that begins, "When I am planning . . ." Where some take the analysis of planning to center on the local state, I take it to focus on political agency, staged by political-economic structure and culture as well (cf. Clavel 1985, Krumholz and Forester 1990).

Some see citizens' participation in planning processes as matters of advocacy and legal rights (Checkoway 1994); I take participation not only to present well-known dangers of manipulation (Arnstein 1969), but also to present real political opportunities for deliberative, even transformative, learning and participatory action research. In a world in which rights are not self-implementing, advocacy planning in many forms is essential, to be sure. Some focus on rationality in planning as a mental process of decision making (Alexander 1996); I take rationality to be an interactive and argumentative process of marshaling evidence and giving reasons, a process that in principle minimizes excluding relevant information and encourages the testing of conjectures, a process that welcomes rather than punishes value inquiry (Fischer and Forester 1993, Gutmann and Thompson 1996).

Where some see pragmatism loosely connected to ethics, more a process of learning from experience (Schön 1983), with Hilary Putnam (1990), I take pragmatism and our pragmatic action to be integrally connected to ethics, for we learn in action not only about what works but about what matters as well. Some see planning as having a dark side (Flyvbjerg 1996); I see planning as having many sides, some controlling and others empowering, some more strategic and others more deliberative—with the structure and politics of planning always shaping, rather than predetermining, the character and consequences of planners' work in any real case.

Some see deliberative, consensus-building processes as relatively insulated from ideology, imbalances of power, and structural political-economic forces (Innes 1996). I see such deliberative processes as precarious and vulnerable achievements created on existing political stages. Where some take mediated negotiations and consensus-building processes to be stand-alone procedures, I take them to be directly related to the ongoing staff duties of planners and policy advisers who must work with task forces, special committees, working groups, and formal and informal advisory boards every week (Susskind and Cruickshank 1987, but cf. the seminal Rivkin 1977). Some see practitioners' reflection-in-action as a largely psychological process of reframing problems, a process of changing one's mind (Schön 1983); I see such re-cognition as integral to deliberation in which parties together learn about fact, value, and strategy all together.

From beginning to end, this book seeks to establish a dialogue of practice and theory, a dialogue of practitioners' insightful accounts and theorists' careful interpretations, to enlarge our planning imagination in the day-to-day work of meetings, negotiations, and briefings. This book presumes that the insightful analysis of planning situations can encourage better practice not by producing abstract lessons but by showing what can be done through practitioners' vivid, instructive, and even moving accounts of their successes and failures.

By examining practitioners' practice stories as windows onto the world of planning, this book radically extends and deepens my earlier *Planning in the Face of Power* (1989), both practically and theoretically. Probing those stories rests on an assumption not widely shared with others in the fields of planning and public policy. Astute practice in messy, highly politicized settings provides important intellectual challenges to planning theory, and to social and political theory more generally. Tacit practice can lead written theory. Skillful practice can help us to identify problems and to refocus practice-oriented theory, whether those problems involve political power or cultural difference, popular mobilization or social justice. Accordingly, this book presents a plea for academic "theorists" to take practice more seriously, to recognize sensitively and to analyze powerfully what insightful practitioners do well in the most challenging moments of their work. Put a bit moralistically, criticism is too often cheap, and planning bashing is a national pastime. The point here is not to celebrate planning, but to encourage its more politically astute and ethically critical practice.

My Bias for Practice

My method rests on three premises. First, these chapters presume that as students of planning and public policy, we should take the messy, conflicted, dirty-hands experience of planners and policy analysts seriously. We should search especially carefully for examples of practice that exemplify what too many others only preach: dealing with differences of values, culture, and ethnicity; dealing with turf problems among agencies, dealing with distrust and suspicion, inherited anger and resentments, and so on. So I have chosen cases from two sources. I have drawn first upon a pool of several hundred interviews devoted to producing "profiles of practitioners"—first-person accounts of the difficulties and strategies of planning practice (see e.g., Forester and Kreiswirth, 1993a, 1993b, 1993c, 1993d; Forester, Pitt, and Welsh, 1993; Forester, Fischler, and Shmueli, 1997). Second, I have drawn on observations and tape recordings of several years of periodic planning staff meetings in a New York State municipality (e.g., Forester 1993a). As a result the following chapters explore the challenges, strategies, traps, and possibilities of planning through actual practice stories drawn from fields of environmental, transportation, land use, housing, economic development planning, and more. When this book suggests insights and when it sketches instructive theory, it does so through the lived experiences of planners and policy analysts, not as a substitute for those experiences.

Second, this book presumes that planners and public policy analysts have to improvise in complex and novel situations all the time. Improvising means doing the best one can in the circumstances at hand, hardly going by the book. The good improvisor has to exercise practical judgment, fitting practical strategies to unique situations to achieve good results. These chapters assess such improvised judgments carefully as practitioners respond for better and worse to the political forces staging their work, as they resist insidious mandates or bury their heads. Along the way, too, I suggest how planners and policy analysts can fail if they ignore the issues explored here.[2]

Third, this book presumes that no realistic discussion of planning and policy analysis is possible without taking power into account in several forms. Planning and public governance obviously take place on stages permeated and structured by relations of political-economic, bureaucratic, and social power, and this book wastes no time prescribing behavior for someone's imagined but wholly unattainable ideal. But precisely because of the pervasive influences of power, this book issues a challenge to students of planning: Let us stop rediscovering that power corrupts, and let's start figuring out what to do about the corruption. Let us not just presume as unshakeable truth that disciplinary power is total, that rationality self-destructs, that hegemonic culture is all pervasive, that we can do nothing to address inequality, poverty, environmental destruction, and needless human suffering. If we are to analyze power as a political, and thus alterable, reality rather than as an unchangeable metaphysical ether, let us stop rediscovering power and instead assess practically, comparatively, and prescriptively what different actors can do about it (Forester 1999a). Let us not only ask but begin to answer, "What ought to be done, when and how, in real cases that matter?"[3]

Precisely because planning and policy analysis take place in a political world, planners and analysts need to anticipate and respond to foreseeable relationships of power and domination. Precisely because severe inequalities of wealth and power, opportunity and victimization persist in cities and communities, we need practically sensitive, politically realistic, and theoretically insightful accounts of democratizing and advocacy practices.

Hardly an ideal form of dialogue, real public deliberation suffers from inequalities of power, poor information, inadequate representation, histories of violence brought to the table, histories that silence the voices of many parties (Beauregard 1989, Guttman and Thompson 1996,

Sandercock 1995). To respond effectively to these typical problems, we need to assess the real and possible practice of public deliberation more closely than we have.[4] This book explores this work from the inside, through practitioners' voices, feeling with them as parties threaten to disrupt processes, as parties voice suspicions and distrust, as parties vent anger and resentment at the history of government action or inaction in their neighborhoods, toward their interests, rolling over their communities.

In a public policy world riddled with racial violence and discrimination, with vast differences in levels of political organization and mobilization, insight and good theory are not luxuries but practical necessities (Beauregard 1995). We need more than ever before the sensitive recognition of differences and needs *and* the thoughtful political construction of practical strategies of response. This explains the value orientation or the normative basis of this book: not any prior assumption of good intentions by planners, but rather a value-explicit inquiry into how to do planning well in a messy, politicized world. I do not assume that good planning should always maximize any one value—equality or autonomy or respect or voice—over others in all cases. I ask instead in a range of cases how a practice fostering public deliberations can succeed or fail: how such deliberative encounters can fuel resignation or empower action, how they can rationalize decisions already made or enhance public learning and stronger democratic practices.

Rather than assume that planners have autonomy, I explore throughout how planners might use discretion and judgment to carve out autonomy, to take risks here, to encourage community mobilization there, to work creatively as mediators here, to negotiate astutely there. I do not begin with a once-and-for-all definition of good planning. I begin instead by asking how public-serving planners can work, for better or worse, when multiple values come into play—values of public learning and public participation, equality and voice, values of efficiently serving need and effectively reducing unnecessary suffering.

So my choice of cases—or planners' accounts of their practice—is both selective and suggestive, quite deliberately and practically biased toward experiences and judgments that point to positive possibilities: positive because they enhance learning, or respect, or mediated agreements that achieve mutual gains, or the working through of past loss, or successful work in racist settings. I have chosen cases precisely because they do not reflect business as usual; they reflect real possibilities of what planning might yet be.

Accordingly the value or normative arguments of this book are developed explicitly all along the way. But I have mixed more with less traditional academic approaches. In several of the chapters, I provide direct commentaries on practitioners' accounts of practice; in others I first work through the literature on the issues I discuss. For the intellectual history of the issues explored in this book, the endnotes provide directions for further work and many references.

Integrating Form and Content: Taking the Book's Message to Heart

Presenting ideas about good planning, or even being prescriptive, does not mean providing a recipe-like list of "thou shalts" with general instructions. Instead this book explores the practice of better or worse deliberation in its form as well as its content, for it tries to show and model promising deliberative practice as well as to describe and analyze it. As the first five chapters present participants' accounts of challenging practice situations, they ask readers to do some of the work that they are reading about. They ask readers to consider and interpret carefully practitioners' stories so they can learn from them, so they can get tips and clues, reminders and cues, as they come to see the messy situations of their own practice in potentially new ways. In this way this book invites readers to try in their own reading to do what good deliberators must do too: to listen and learn, probe and explore, sort through ambiguity to clarify or recognize underlying value.

But by asking readers to work through the commentaries on practitioners' stories, this book asks those readers to take multiple perspectives—to listen to several voices and then to mediate between them, keeping several claims in contention and searching for practical implications. As chapters 1 and 2 argue, learning to listen and see in this way develops the capacity that scholars call "political judgment" (Sager 1994a).

The deliberative practitioner learns from conversation and argument, the actual interpretation and reconstruction of what parties working together say and do. In deliberative practice, critical listening, reflection-in-action, and constructive argument all interact. The practicing planner or policy adviser must not only think with others, but often must think ahead of time, "How can I put my case so that others will not misunderstand, so that we can move ahead? What should I *not* say? What can I suggest or ask that will help us along?" [5]

This is no simple political work. When planners and policy advisers hope to encourage productive and well-informed deliberative discussions—at a city council meeting, a local planning board, a state legislative subcommittee—they need to anticipate the plural and conflicting stories of differently affected citizens and stakeholders. Chapters 1 and 2 show that such anticipation requires imagination and emotional responsiveness, the capacities to empathize with other parties *and* to remain politically critical at the same time. Working among conflicting parties, planners and policy advisers must be able to recognize in detail the perspectives of others, their stories and accounts, their feelings and stakes, without necessarily agreeing with any of them. As chapters 3 and 4 go on to show, the challenges of fostering public deliberation require both closeness and distance, empathy and critical judgment, and the character to recognize and respect, rather than to dismiss, the human emotions of anger and fear, impatience and suspicion. Encouraging participatory processes requires as well a commitment to shared evidence and good argument, a commitment to the distinctions between warranted truth and demogogic posturing as well.[6]

In practice, planners and policy advisers must do much more than rehearse public deliberations imaginatively before the fact. They must make them work. They must convene and staff public meetings, provide briefings to participants, respond to the needs of several parties at once, and many times shuttle back and forth to meet with angry and conflicting parties. The resulting public deliberations are iffy and contingent, precarious and vulnerable, but planners can play mediating midwifery roles nevertheless. Often planners must bring conflicting claimants to and through the public arena to deliberate practically together: to participate together and learn, to reach joint gains whenever possible, to craft effective strategies and real options, to implement and to meet their needs—not just to encourage deal making behind closed doors, not to cool out angry publics, not to minimize participation to satisfy meeting requirements, not just to maximize noise at pro forma public hearings.

This book explores strategies that practitioners can use to encourage fruitful and participatory public deliberations, strategies of community planning, participatory action research, and mediated negotiations (cf. Greenwood and Levin 1999; Forester 1989: chap. 6).

Listening, Learning, and Implications for Practice

The practitioners' accounts that follow do not conclude with generalized hypotheses, but they are likely to be more, not less, useful as a result. Integrating politics and history, challenging any simplistic oppositions between rationality and feeling, intelligence and sensitivity, these practice stories help us to identify the personal challenges of deliberative planning and policy analysis. We can learn from these planners' accounts about hope and discovery, about surprise and coming to see our problems anew. We learn about capacity building in organizations and planning processes, about relationships that can be built from shaky foundations of suspicion and distrust. When East St. Louis is so poverty ridden that a court has ordered the city to sell off city hall, no individual's planning strategy will overcome that blight. When infested mudwalls of campesino housing pose public health threats, no planner's magic bullet will provide freedom from disease. But even if individuals cannot accomplish what social movements for jobs and health, housing and equality, might, that is hardly reason not to work to create jobs, to improve housing and health, to plan where effective public and community action may still be possible. In cases like these, not only can the odds of mitigating poverty seem overwhelming, but so can the extent of need and suffering. What can really be done? How can plural and conflicting parties work to rebuild their communities together?

By exploring planners' own stories, the following chapters help us to begin to answer these questions. These stories can teach us about a critical pragmatism, about risk taking, searching and innovating, about resisting the cynicism of putting aside our ideals, about the willingness to do hard work in messy circumstances to repair the world, today and tomorrow. These chapters focus on performance and not trends, on politically staged action and not general strategies.

Assessing these planners' stories, we find the political drama of planning to be captivating and instructive because the scripts are only intimated, not written; the political actors are creating their future, their lives, as they act with one another, searching for rewarding steps ahead, not abstract ideals. But the drama of planning and public policy is tragic too—not only because the few rich dominate the many poor, because the wealthy have several homes and the poor are landless, but for a more profound, if ordinary reason as well. As individuals or as communities, our values conflict and cannot all be realized. This is the meaning of political plurality—our caring about many values. This is not just an illusion of capitalist-induced scarcity. But tragic conflicts of value (obligations to different generations or equal parties within one generation) do not diminish by one iota real human needs for food or dignity, autonomy or community. That our grappling with public choice and democratic politics is real, ongoing tragic drama on the public stage justifies not a twinge of needless suffering, not a moment's disregard of joblessness, ill health, poor housing, sexism, or racism. The dramas of building and rebuilding our cities unfold before us as we act, set on stages made by broad historical forces of class and technology, culture and politics. This book explores the action in those dramas by listening to and learning from actors who share our stage.

Readers will find characters here with whom they can and cannot identify, and both experiences are important. The power of the planners' stories flows in part from their ordinariness. "Yes," we say, "I've been in just that situation. I wonder how this person handled it?" But that power flows too from the sensitivity and guts of the planners who speak here, and it will be the rare reader who does not find himself or herself moved by these accounts and inspired by the perceptive and effective practice displayed here.

Such inspiration is hardly a social-scientific failing of the book, though.[7] On the contrary, such inspiration is likely to be important when connected to the objective circumstances and insights that enable effective practice. Inspiration that manipulates hope and vanishes is destructive, distracting, and perhaps soothing, but pragmatically worse than useless. Inspiration that fires the imagination and trains perception, that encourages listening and keeps hope alive—that inspiration flows from courageous practice against the odds, from pluck and persistence, from the creativity and finesse demonstrated in many of the accounts that follow.

The practitioners' stories in this book are literary and pragmatic at once. These stories illuminate complex and messy situations of real life no less than they portray the tragic choices citizens face in a world of deep conflict. These stories reveal the political and cultural dramas played out in the actions of city building, community development, environmental protection, historic preservation, and so on.[8] When citizens have a lot to lose or gain, when reputations are at stake, when the economic livelihood of neighborhoods is at stake, the work of actual planning and policy analysis is real drama—drama no less powerful, no less moving, no less instructive, and no less illuminating for being set on the stages of our cities and neighborhoods.[9]

I

Deliberative Practice Reconciles Pragmatism and Vision

1

Listen to Stories, Learn in Practice: The Priority of Practical Judgment

In an insightful book entitled, aptly, *Not Well Advised*, Peter Szanton explored the problems of linking the research capacities of universities to the needs of our cities.[1] Making those linkages work had been very tough, and Szanton wanted to explain what had happened, what was not workable, and what might yet work. In the closing chapter, "What Have We Learned?" we find the warning, " 'Generalizability' is a Trap," and the striking lines, "F. Scott Fitzgerald commented—on the writing of fiction—that if he began with an individual, he soon had a type, but if he began with a type, he soon had nothing." Szanton went on to argue that a similar rule held for social and political analysis hoping to make a difference, for applied social research.[2] In recent years, several other students of planning and policy analysis have been exploring closely related themes, and their arguments are practically pitched and quite instructive.

In a popular book on the uses of history in policy analysis, *Thinking in Time*, for example, Richard Neustadt and Ernest May recommended a practical maxim they called the "Goldberg Rule" (Neustadt and May 1986:274, 106). They told planning and policy analysts, "Don't ask, 'What's the problem?' Ask, 'What's the story?'—That way you'll find out what the problem really is."

In planning, Martin Krieger's *Advice and Planning* began to explore the importance of stories as elements of policy advice, and Seymour Mandelbaum, James Throgmorton, Leonie Sandercock, and Howell Baum have argued that our stories define us in subtle political and social ways, expressing and reshaping who we are, individually and together (Baum 1997, Krieger 1981, Mandelbaum 1991, Marris 1990, Sandercock 1995, Throgmorton 1996). And in political theory and philosophy, Peter Euben and Martha Nussbaum have argued that literature and drama, and tragedy most of all, can teach us about action, ethics, and politics in ways that more traditional analytic writing cannot (Euben 1990, Nussbaum 1986, 1990a).

On Nussbaum's account, for example, literature can give us a fine and responsive appreciation of the particulars that matter practically in our lives; literature, she suggests, can give us an astutely alert pragmatism, hope, with less false hope, a keener perception of what is really at stake in our practice—in effect, a realism with less presumptuousness about clean and painless technical or scientific solutions—be those the solutions of either the hidden hand of the market or the more visible fist of the class struggle.

The broader practical relevance of these writers' concerns was captured wonderfully by Robert Coles in his book *The Call of Stories*. In the opening chapter, "Stories and Theories," Coles provides a moving and resonant account of his early clinical, practical training in psychiatry. Coles introduces us to two of his supervisors: Dr. Binger, the brilliant theorist who sought out "the nature of phobias," "the psychodynamics at work here," and "therapeutic strategies," and Dr. Lüdwig, who kept urging Coles to resist the "rush to interpretation"—a wonderful and poignant phrase!—by listening closely to his patients' stories.

Coles writes that for thirty years he has heard the echoes of Dr. Lüdwig's words: "The people who come to see us bring us their stories. They hope they tell them well enough so that we understand the truth of their lives. They hope we know how to interpret their stories correctly. We have to remember, that what we hear, is their story" (Coles 1989:7).

The important point here is *not* that psychiatric patients have stories, as we all do, but that Dr. Lüdwig was giving young Dr. Coles some practical advice about how to listen, about how much to listen for, and about the blinding dangers of rushing in "theory first " and missing lots of the action. Dr. Lüdwig was giving Coles, and Coles is passing along to us, practical advice about learning on the job—about the ways our current theories focus our attention very selectively, as a shorthand perhaps, but if we are not very careful, too selectively.

Coles recalls his mentor, physician and poet William Carlos Williams on this danger of theoretical oversimplification. Williams wrote, "Who's against shorthand? No one I know. Who wants to be shortchanged? No one I know" (Coles 1989:29). When we have practical bets to make about what to do and what might work, theory matters—but so do the particulars of the situations we are in, if we want those bets to be good ones, if we want not to shortchange others or ourselves.

Stories matter practically in daily work and in the professional school classroom too. In an undergraduate course, "Planning, Power, and Decision Making," recently my students read, as they typically do, a mix of planning case histories, interwoven with historical and theoretical material about power and powerlessness. But for one week in this class, I tried something new. I asked the students who might be interested to read thirty to forty pages of edited interviews with practicing planners who had graduated from our program (Forester and Chu, 1990). These edited interviews took the form that Studs Terkel has used in his many books: we removed the interviewer's questions so the planners' stories remained—the planners' own accounts of the difficulties, surprises, rewards, and frustrations of their real work. What happened? The result in class was striking:

"Now I can tell my mom what planners really do. It's not all one thing, but *this* is it! Now I can tell her!" Another student said, "This was the most practical thing I've read in three years in this program!" (As the professor, frankly, I did not know whether to laugh or to cry. Much of what I had assigned for the previous ten weeks, I thought, including case studies and a negotiation primer, had been pretty practical!)

Several other students had similar reactions. Somehow the profiles had grabbed them in eye-opening and obviously effective ways. Why had that happened? What was so striking, so catching, and so effective about those stories told by planners about their own work?

The stories were certainly concrete and descriptive, hardly marred by abstract, theoretical, or unfamiliar language, but that was no explanation. The same was true of the historical material in the required readings. Something much more important than concreteness was at work here, with implications that reached far beyond the classroom.

Over the next several weeks, I began to think of this classroom experience in the light of recent studies of policy analysts and planners. Typically that work involved not only interviews with planners and analysts, but it also grew out of observations of formal and informal meetings in policy and planning processes, including, for example, planners' and analysts' own staff meetings (Baum 1990, Feldman 1989, Forester 1989). Beginning to assess the politics of analysts' and planners' speech—and their listening more or less well to others' speech too—this literature suggests that practicing planners both tell and listen to practice stories all the time (Healey 1997, Throgmorton 1996).

Would it be very surprising, then, if analysts and planners at work learned from other people's stories just as the students read, listened to, and learned from the practitioners' stories they had read in class? Hardly, and taking this question seriously can lead us—as it has led me—to participate in planning and policy meetings more effectively and to listen more insightfully to the stories that planners tell and hear as they work with others. Three questions about these practice stories quickly arise. First, what do planners and analysts accomplish as they tell such practical stories at work? Second, what kinds of learning from such storytelling are possible and plausible? And third, what does such storytelling have to do with the politics of planning and policy analysis?

The following sections explore these questions. First, we turn to an excerpt from a profile presented in the classroom. Second, we look closely at a segment of an actual city planning staff meeting—a segment in which the planning staff listen critically to, and reconstruct, a practice story told by the planning director. Third, we consider several ways we can learn from practice stories on the job (and more broadly as well).

Learning from Practice Profiles

How do we learn from the practice stories told by practitioners in their own voices? The following selection comes from an interview profile of "Kristin" (not her actual name), a recent professional school graduate (Forester and Chu 1990:1–8). Kristin has been describing a lengthy process of meetings she had held with residents and commercial interests in a neighborhood to discuss zoning issues—to allow concentrated, possibly mixed-use, commercial development and to prevent residential displacement too. Out of the process came a proposal that went to the city council, twice. Kristin put it this way:

At the second meeting, there was movement toward an agreement, but a councilman made a motion to drop the height by thirty feet in all of the areas, and it undid the whole thing. It upset the balance we had worked out. The developers jumped to their feet and rushed to the microphones and said, "Look, we worked with the planning staff long and hard to determine these heights, and they're not just drawn out of thin air. They're related to densities and uses, and these are the numbers that work. You can't just go in and chop!"

Fortunately, the council listened to the voice of reason and they agreed, but very hesitantly. They didn't want to, but they agreed, since it was only a land use plan and wasn't the actual zoning itself. Since the zoning was to be decided later, the council figured, "If we want to change our minds, we'll do it then." And they let the plan go through.

We'd built a consensus, and it really was very fragile, because different groups had very different ideas from the start. When we went to city council for the second time, we thought, "Okay, now we have it, because now everyone's happy, and we're sure everyone's happy." Then the process breaks down a second time because of this idea of heights. The people who were really affected were satisfied, but because of one voice, the process stopped. It was just a lot of delay, a lot of frustration, and a lot of uncertainty. I really thought the whole thing was going to just come apart.

And I hate to say this, because it sounds terrible, but in the end it almost doesn't matter. When our office goes to the Planning Commission, we go with staff recommendations. Although we take into consideration a lot of what goes on, in the end we don't really need to have a consensus because we only present the staff position. It's *up to the citizens* and the developers and other interests to come and present their own perspectives to the commission. In this city, our staff position doesn't carry the greatest weight. We can work on something for a very long time, but if someone comes into a Planning Commission meeting and makes a statement, they'll just undo what we've spent months working at. If we take long, hard months and go through the process of building consensus, in the end it almost doesn't matter. I don't think the Planning Department's work has a lot of clout.

I've been very disillusioned. Of course people are going to be self-interested, but I was surprised at the degree to which that's true, and at how people work very long and very hard simply to protect their own interests. The city council is just an extremely political place. This is the first planning job I've had, and I often find myself wondering, "Is this how it is, or is this how it is here?" Most of the planners I talk to are also fresh out of school and they say the same thing.

Still, there are little successes. We just relocated a government facility that the government provided and expanded, and it didn't really stand in the way of our plans for the area to try and transform it. We got in on time on that and helped to get them a place that was more appropriate, and everyone sort of won. It's those little things that make me feel that maybe in the long run I can make a difference. In the long run, it's the little day-to-day things that come up that give you a chance to make a neighborhood a more pleasant place to live in. The small things make up for other things. I say, "Okay, I'll put up with certain things because that's the price I have to pay to be able to do the other small things that *do* have an effect."

It's also been a learning experience. This is my first job and even though sometimes I'm not real happy with the way things go, I'm still learning a lot about politics." (Forester and Chu 1990:63)

How do we learn from such stories, and what do we learn? From Kristin's account, we learn much more than the simple facts of a case. We learn, first of all, a good deal about her: her disillusionment and her sense of satisfaction with the "little successes." We learn about her expectations and her awakening to the politics of the planning process—and as readers we are obviously invited to compare our expectations of the process with hers, perhaps to be awakened in the same way (Nussbaum 1990a:148–167).

We learn about the vulnerability of planners' efforts: "We can work on something for a very long time" and "someone" can "just undo what we've spent months working at." We learn about a plausible, if uneasy, view of planners' roles ("Although we take into consideration a lot of what goes on, in the end we don't really need to have a consensus because we only present the staff position" to the Planning Commission)—and we learn about what this view of her role implies about the encompassing politics of planning. So she says, "It's up to the citizens and the developers and other interests to come and present their own perspectives," and having expressed this view, Kristin wonders if, and doubts that, the Planning Department's work "has a lot of clout."

And there is more, from insights about timing and politics ("Since the zoning was to be decided later, the council figured, 'If we want to change it, we'll do it then' ") to Kristin's own sense of realism and hope ("This is my first job, and even though sometimes I'm not real happy with the way things go, I'm still learning a lot about politics"). That's learning about politics that Kristin's story shares with us, so we can learn too. In the face of power, she suggests, planners must pick their targets carefully.

Yet many of us are suspicious of such stories. What can we learn from them that is not simply unique and idiosyncratic? For years, the distinguished practitioner-turned-academic Norman Krumholz doubted the value of writing up his ten years of experience as the city of Cleveland's planning director, trying to make equity planning work. The urgings and encouragement of his many colleagues and friends notwithstanding, Krumholz suspected that his and his staff's experience would be too unique, too particular, too much "just about Cleveland," for others in other cities and towns to learn much from, or to find really relevant. Those stories, though, are careful accounts of the complexity of big city planning practice, and they turned out to be vivid, instructive, sobering, and inspiring all at once (Krumholz and Forester 1990).

Krumholz's suspicions had been fueled by an academic culture that judges any work not conforming to canons of systematic "social science" as guilty before proven innocent. "Physics envy" in social research is alive and all too well. The point here is not to scapegoat positivism or modernism, but to recognize that the imperialism of a narrowly construed social "science" has often terrorized graduate studies and social inquiry more generally. We forget too easily that science is a cultural form of argument, not a valueless, passionless use of magical techniques (Gusfield 1981; McCloskey 1985; Beauregard 1989, 1991, Fay 1996). Anthropology, for example, is a social science from which few would doubt that we can learn, and we would be silly to dismiss anthropology—and perhaps the field of history too—because it is not a "scientific" discipline in the experimental, culturally conventional sense.[3] The point here is not to argue against hypothesis testing when it is possible, not even now to argue for a desperately needed, broader conception of social research, but to pursue the question of how practitioners learn and develop good judgment in practice—especially in applied and professional fields like the design and policy-related professions.

In practice, the real-time demands of work allow for little systematic experimentation. Just as clearly, practitioners at work engage in what we might call practical storytelling all the time, telling, for example: what happened last night at the meeting, what Smith said and did when Jones said what she said, what the budget committee chair did when the citizens' action group protested the latest delay, and what happened with that developer's architect's last project. In practice situations we find stories and more stories, told all the time and interpreted all the time, sometimes well, sometimes poorly, but we find relatively few "controlled experiments."[4]

We're likely to find far more stories, too, in practice settings than we will find opportunities to try things out, to test our bets, to move and reflect-in-action, as Donald Schön has so powerfully described it in *The Reflective Practitioner* (1983). Faced with such stories and paying careful attention to them, planners and policy analysts learn in practice about the fluid and conflictual, deeply political and always surprising world they are in. Listening for far more than words to assess the organizational and institutional contexts at hand (Forester 1989:chap. 7), planners and policy analysts, we shall see, must often be not only reflective practitioners acting by themselves, but deliberative practitioners acting with others as well.

But how do they do it? What does it mean to pay such careful attention to these stories and so learn from them? How can we explain and dignify the ways planners and analysts can learn practically and politically, as they listen to these stories?

Practice Stories Told and Listened To in Practice

To explore how practice stories might work in actual planning and policy settings, we can sit in, for example, on a city planning agency's staff meeting. Here we see not only that the professional planners tell one another practical, and practically significant, stories all the time, but also that they are creating common and deliberative stories together—stories about what is relevant to their purpose, stories about their shared responsibilities, about what they will and will not, can and cannot, do, about what they have and have not done.

To see how such common and deliberative stories can work, we turn to a brief excerpt from a planners' staff meeting in a small city. We can think of listening in on a staff meeting as a way of getting inside the organizational mind of the planners, getting to know both how they perceive the situations they are in and how they begin to act on the problems they face.

In this case, the staff number roughly half a dozen professionals. The meeting followed a recent election in which the mayoral challenger, who lost narrowly, had run a campaign vigorously attacking the successful incumbent's planners—the planners holding this meeting. As the transcript suggests, the staff feel, to say the least, unappreciated and hardly well understood by the public. This segment of their conversation begins as follows:

The director remarks: "I think the Mayflower project [an apartment complex] was pivotal. That's the first time that we took a very high profile position on a very unpopular issue. We were outvoted on council 7–3; we were pushed right to the center of that controversy. We tried to hold what we thought was the right line, and we really lost a great deal of support in the general public because of our position.

"I think that was the first real bad one. And it gets blended in with the Northside. I think the Northside's the second one where we've been hurt, where those people who are afraid and concerned are really, really angry. And Lakeview Park's another one, although I don't think we're taking the heat for that one."

"I think we are," says George, the community development planner.

"I think we are too," adds Karen, the housing planner.

"I mean," says the assistant director, "I don't hear anybody saying, 'The Board of Public Works really screwed up.' "

The director replies, "It's interesting, because it wasn't our screw-up."

"But," says Karen, "*we never said* it wasn't our screw-up. We never pointed the finger to the screw-up."

The director responds again, "My perception is that people just think about any kind of change, and then they think about planning, and then they think about planners, and . . . somehow we're tied to everything."

Bill, the senior planner, agrees, "It's guilt by association."

Karen expands the point: "If something goes wrong, the planners did it. If something goes right, the city council members claim credit for it."

And the director replies, "*That's* the kind of problem I think we have to address. Miss Smith here has been saying, 'You've got to come out,' and you [have] too, and many of you have been saying, 'We've got to answer this, we've got to answer this.' And I've always said, 'No, we *don't*—because we don't want to get into a cursing war with a skunk; you know, you just get more heat that way.'

"But now I think *we have to*. I think we have to set out a strategy over the next year or two of how we're going to sell the department and how we're going to position ourselves to get to those people whose minds aren't already made up. I mean, you'll never get Samuels [a local journalist] to think we're good guys, but there are a lot of people influenced by Samuels and the crap he's saying. If we could get to them with reason and explain to them what our job is and how we came to the conclusions that we've come to . . ."

Here George, the community development planner, suggests, "What about the concept of developing something like a position paper for the individual projects, like the Mayflower project that would be very much a synopsis, but at least it would state when it came to the department, what the developers' request was, what our recommendation was, what the council did, you know, a 'who struck whom' sort of thing, and what the key issues were for the neighborhood, and how it turned out?"

And the staff here go on to discuss these issues.

Quite a bit is happening in this working conversation. The director tells a story about their efforts—what they have been up against, what they have tried to do. It has a time line; the Mayflower project was "pivotal"; "that was the first real bad one," he claims. And he connects that experience to others—the Northside and the Lakeview projects. And he does more. He characterizes the people involved: they're "afraid," "concerned," and, he says, "really, really angry." He does not just describe behavior, but he socially constructs selves, reputations: the kind of people—he is claiming—whom the planners have to work with, "really angry."

The director is doing much more than that too, for he tells a complex story, in just a few lines, about the allocation of responsibility and blame: "We tried to hold what we thought was the right line"; "We took a very high profile position on a very unpopular issue"; "We really lost a great deal of support," but on Lakeview, "I don't think we're taking the heat." But when two of his staff think instead that they *are* "taking the heat" for it, the director says, "It's interesting, because it wasn't our screw-up." So a story unfolds here about the courage of convictions, about the tension between commitment to a professional analysis and the desire for public support, about astute or poorly played politics, and a story about guilt by association—the vulnerability of the planning staff in a highly politicized environment.

This conversation begins with a working story of effective and vulnerable practice, practice that is strategic and "contingent" (as planning professors and consultants now say), and the conversation includes the retelling of this story, developing the story so that there is a moral: a lesson and a point, a clarification of the situation the planners are in, and a clarification of what they can now do differently and better as a result.

The director has spoken the most here, but he does not just tell a story to an audience. The planners together—with differences in their positions, power, and influence, to be sure—work to develop their own story, for it is, after all, the story they are willing and practically able to construct together.

Karen echoes a pervasive problem in planning when she focuses on a particular irony of their practice: "If something goes wrong, the planners did it. If something goes right, the city council members claim credit for it," and the director responds to her, affirming and building on her moral claim about the allocation of credit and blame: "That's the kind of problem . . . we have to address."

So he proceeds to reconstruct and present their working history again: "Miss Smith here has been saying, 'You've got to come out [more publicly],' " and "Many of you have been saying, 'We've got to answer this,' but "I've always said, 'No, . . . we don't want to get into a cursing war with a skunk.' "

Then he tells the staff he has changed his mind, so their collective story is changing: "But [now] I think we have to. We have to set out a strategy . . . to sell the department . . . to get to those people whose minds aren't already made up."

The community development planner does not then just take that as a cute personal tale of the director's change of heart. He takes it as a working story about where they practically are as a staff, so he brings up a strategy to be considered: "What about . . . developing something like a position paper for the individual projects?" He goes on to sketch his idea for the staff's consideration, for their deliberation.

What we see here, even in this short stretch of conversation from a staff meeting, is very rich, morally thick, politically engaged, and organizationally practical storytelling. The point is not that planners tell stories, for everyone tells stories. Rather, in planning practice, these stories do particular kinds of work—descriptive work of reportage, moral work of constructing character and reputation (of oneself and others), political work of identifying friends and foes, interests and needs, and the play of power in support and opposition, and, most important (here in the staff meeting, for example), deliberative work of considering means and ends, values and options, what is relevant and significant, what is possible and what matters, all together. The staff do not assess means and strategies alone, as if values and ends were just given, to be presumed. The staff try to explore and formulate what matters and what is doable too.

Most important, these stories are not just idle talk; they do work. They do work by organizing attention, practically and politically, not only to the facts at hand but to why the facts at hand matter.

In any serious staff meeting, for example, these stories are ethically loaded through and through. They arguably ought to be "relevant," "realistic," "sensitive" to the staff's political history, respectful of important values at stake, always alert to the idiosyncratic wishes and strong feelings of other public officials, community residents, planning board, and city council members alike. Carefully telling these practice stories and listening perceptively to them are essential to the planners' work of astutely getting a fix on the problems they face.

In their meetings, the planners do not simply tell individual tales. They work together (as they work with others in other meetings) to construct politically shaped, shared working accounts, commonly considered, deliberative stories of the tasks, situations, and opportunities at hand. In these deliberative stories, the planners not only present facts and express opinions and emotions; they also reconstruct selectively what the problems at hand really are. And they characterize themselves (and others) too—as willing to act in certain ways or not, as concerned with these issues, if not so much with those, as having good or poor working relationships with particular others, and so on.[5] We might say that not only do we tell stories at work, but our stories tell a good deal about who we are as well.

We began by exploring the ways we learn from profiles of planners—planners' accounts of their own practice. But that problem of learning from practice stories now appears to be widely shared with professionals of all kinds who must listen to and interpret the stories they hear from their patients and clients: the developer wanting to build, the neighbor wanting to protect the neighborhood, the politician wanting some action from the planning department, the planning board member asking why more has not been done (and done more quickly) on a given project. What professionals generally and planners particularly face in such cases are complex stories, and if they do not learn from them, and learn quickly, they are likely to find themselves in serious trouble.

But we still have not answered a central question: How do planning analysts learn from such practice stories, and how do they learn politically and practically from such storytelling if they obviously do not do it through systematic experimentation? How do planners and policy analysts learn from other people's practice stories when they cannot do much hypothesis testing on the spot? As Donald Schön's analysis of the moves that enable reflection-in-action shows, we do some hypothesis testing as we act, but we seem to learn a good deal more, and to reflect on a good deal more as well, than Schön's account encompasses (Schön 1983; compare chapter 5 below). So what image of practical and political learning can provide us with a fresh view of these questions, if the imagery of learning through "experimentation" is not really apt in practice?

Learning on the Job as We Do from the Insights of Friends

Recall first that when we ask how planners and professionals learn on the job, we are asking how they come to make practical judgments when time, data, and resources are all scarce. We are not asking how "scientific" professionals can be; we are not asking how well they remember certain methods courses they took in school. We are asking how they learn in the thick of things, in the face of conflict, having to respond soon—not how they learn through sustained research.

Second, we should remember that when professionals learn on the job, they have to make judgments of value all the time. They not only have to make value judgments "in their heads," but to make value allocations as they speak, as they make practical claims about what is to be seen as working well or poorly, about who is to be regarded as reliable or not, about what is to be taken as important, noteworthy, worth time and attention, and what is not by their listeners (or, when they write, by their readers). They must learn not only about what someone has said, but about what that means, why that is important, and why that is significant in the light of the inevitably ambiguous mandates they serve, and in the light of the many ambiguous hopes and needs of local residents.

With little time and facing the multiple and conflicting goals, interests, and needs of the populace and their more formal clients, planners have to set priorities, not only in their work programs but every time they listen to others. They cannot get all the facts, so they have to search for the facts they feel matter, the facts they judge to be significant and valuable. So whether they like it or not, they are practical ethicists; their jobs demand that they make ethical judgments—judgments of good and bad, more valued and less valued, more significant and less—continually as they work (see especially chapter 8). Ethical judgments, however embarrassingly little they may be discussed in planning and public policy schools, are nevertheless inescapable and ever present in practice. Really value-free professional work would be literally what what it says: value free, worthless, without worth.

But what image, then, can help us to understand how planners learn practically, politically, and ethically as they listen to the stories they hear? Perhaps the simplest way to suggest an answer here is to notice that we learn not just from scientific inquiry but from the probing insights of friends as well.

We learn from friends—as Aristotle implies in his *Ethics*—and we need to probe, here, our intuitions about friends and friendship to explore how we can learn from planners' stories, and how we can learn, as planners and as practitioners on the job when we work with others, when we listen carefully to others, paying careful attention to their stories (Sandel 1982: 181, Murdoch 1970, Nussbaum 1990a).

The point here is not that planners are, can be, or should be intimate friends with everyone they work with. The point is rather that if it is not clear how we learn from stories in practice settings, then we should think about the many ways we can and really do at times learn from friends. Five related points deserve attention.

First, we learn from friends not because they report the results of controlled experiments to us, but because they tell us appropriate stories—stories designed to matter to us. "Appropriate stories" are not "appropriate" in some ideal sense; they are appropriate to us and to the situations we are really in, insofar as our friends can bring their knowledge, empathy, thoughtfulness,[6] and insight to bear on our particular situation, needs, and possibilities. When we go to a friend with something important, we expect that friend not to respond with small talk and babble, but with words and deeds, little stories (sharing, confirming, reminding, consoling, perhaps encouraging) that can help us to understand practically and politically what is (and is not) in our power to do.[7] These stories are typically narrative and particularized, not formal, logical proofs—however argumentative our friends may be. So for example the planning director quoted above tells the story not only of his having wanted to avoid "a cursing war with a skunk" but now of his change of mind—a story directly appropriate to and responding to the staff's problem of vulnerability.

Second, we learn from friends because we take their words to help us to see our own interests, cares, and commitments in new ways as we may come to reconsider, for example, how we rank our interests (cf. chapters 5 and 6 below; Lindblom 1990, Mansbridge 1988, Michelman 1988). They help us to understand not just how the world works, but how *we* work, how we are, who we are—including, importantly, what sorts of things matter to us (Fishkin 1991, Taylor 1989). They help us to understand not only how we feel but what we value, not only "where we're at" in the moment but how we are vulnerable, dependent, connected, haunted, attached, guilty, esteemed, or perhaps loved, and so in many ways how we are related to the world not simply physically but significantly, in ways that matter to us and to others.[8]

We look in part, too, to friends to be critical (in ways we can respond to), to "think for themselves" as well as for us too, not simply to condone or agree to our every crazy or troubled or rash or ill considered idea.[9] In the staff meeting Karen does not let the director off the hook about getting the "heat" for the Lakeview project even though, he thinks, "it wasn't our screw-up." She points to the staff's own responsibility: "But *we never said* it wasn't our screw-up. We never pointed the finger to the screw-up"—whether or not "pointing the finger" would have been effective.

Third, we learn from friends because they do not typically offer us simplistic cure-alls or technical fixes. They do not explain away, but rather try to do justice to the complexities we face. They do not reduce those complexities to trite formulas; they do not make false promises and sell us gimmicks, even if they might encourage us and perhaps not tell us everything all at once about what they think we are getting into—if they are confident, for example, that we will do what needs to be done once we get going. Friends recognize complexity, but as pragmatists concerned with our lives, our practice; they neither paralyze us with detail nor hide details from us when they know they will matter.

The lesson here is not that situations determine actions, but that practical rationality depends far less on formulas or recipes than on a keen grasp of the particulars seen in the light of more general principles and goals. Taking practice stories more seriously, we ironically make decision making less central to practice and make the prior acts of problem construction, agenda setting, and norm setting more important (Murdoch 1970:37).

But if they do not offer us technical fixes, what *do* friends offer us? Certainly not the detached advice of experts, for they do not typically invoke specialized knowledge and so tell us what to do. Instead they help us to see more clearly, to remember what we need to, to see in new ways, perhaps to appreciate aspects of others, or ourselves, or our political situations, to which we have been blind (Nussbaum 1990a:160, Wittgenstein 1967:227e). In the staff meeting, Karen's moral tale is poignant and powerful, capturing part of the bind the staff is in (and setting up the director's response): "If something goes wrong, the planners did it. If something goes right, the city council members claim credit for it."

Karen's insight teaches us about the complex rhetoric of democratic politics and participation, its ideals and its ironies. Listening here, we might be less wishful but more astute, less rosy-eyed but more committed to doing what we can. We can gather from these stories too the differences between better and worse deliberation, between more and less inclusive participation, between a more or less "fragile consensus" (as Kristin had put it). Perhaps we learn about the differences between dominated talk and "real talk," and its contingencies (Belenchy et al. 1986: 144–146).

Listening to Kristin, for example, we can understand more about how power and rationality interact, about how what seems well founded may never come to pass, and about how planners' and citizens' good ideas can be watered down, lost in a bureaucracy, held hostage to one politician's campaign. These stories might nurture a critical understanding by illuminating not only the dance of the rational and the idiosyncratic, but also the particular values being suppressed through the euphemisms, the rationalizations, the political theories and "truths" of the powerful (Habermas 1979, 1984, Flyvbjerg 1998, Foucault 1980).

Fourth, we learn from friends because they help us to deliberate (Bernstein 1983, Beiner 1983, Benhabib 1988, Kirp et al. 1989:20). When we are stuck, we often turn to friends if we can. When we need to sort out what really matters to us, we turn to friends—close at hand or even in our imaginations. We look to friends to remind us of what matters, of commitments we have lost touch with, of things we are forgetting in the heat or the pain of the moment. We learn about our relevant history and our future possibilities of practice. We learn too about better and worse (so, without calling it that, about ethics), as we consider our friends' judgments about how to act on our more general goals, in the particular and often surprising situations we face. And enriching our capacity for deliberation is part of what the profiles of planners do, part of what practice stories on the job do.[10] In the staff meeting, George does not let the director stew in his own juices; he suggests a strategy for the staff to discuss and consider: "What about . . . a position paper . . . for projects . . . [stating] when [the project] come to the department, what the developer's request was, what our recommendation was, what the council did," and so on. These practice stories provide empathetic examples as well as abstract arguments about what ought to be done. They allow us to learn from performance as well as from propositions. As Iris Murdoch (1970: 31) put it so powerfully: "Where virtue is concerned, we often apprehend more than we understand and *we grow by looking*."

Fifth, and finally, we learn from these practice stories as we do from friends because they present us with a world of experience and passion, of affect and emotion that, with few exceptions (Baum 1987, Hoch 1988), previous accounts of planning practice have largely ignored. These stories ask us to consider not only the consequential outcomes of planning, not only the general principles of planning practice, but the demands, the vulnerable and precarious virtues required of a politically attentive, participatory professional practice. We should consider applying the Foucaultian move of restoring, resurrecting, or even reconstructing subjugated accounts not only of suppressed, marginalized and dominated groups, but of ordinary planners seeking to attend to issues of public welfare and injustice, need and suffering (Foucault 1980). These stories enrich our critical understanding if they allow us to talk about the "political passions of planning"—the academic undiscussables of fear and courage, outrage and resolve, hope and cynicism too, as planners and other professionals must live with them, face them, work with them.

These stories—whether profiles of planners or practice stories told on the job—engage our emotions and passions, allowing us to learn through whatever emotional sensitivity we have. These accounts help us to consider "how I might have felt in that situation," to explore feelings we might not have recognized as relevant. They develop our repertoires of emotional responsiveness and attentiveness. They teach us through empathy and identification. We learn about situations and selves—our selves—as we imagine being in the situations presented, as we ask, "Would or could I have done that? What should I have done?"

If we could not talk about such political passions, how could we ever talk about any critical practice at all? We would be left with passionless fictions of "correct politics," fantasies either of smooth incrementalist bargaining or an above-it-all problem solving that might inspire illusions of rational control but would hardly be true to anyone's experience. These politically passion-less accounts of planning practice might be soothing and promise a lot, but they would hardly inspire any confidence and hope about the challenges of planning, today and always, in the face of power.

How then can we learn from practice stories? We can learn practically from such stories in many of the ways that we learn practically from friends. Both help us to see anew our practical situations and our possibilities, our interests and our values, our passions and our working bets about what we should do.

The argument here is hardly post-Californian, skeptics should recognize, for the notion of friendship lay close to the heart of Aristotle's *Ethics*. Aristotle distinguished several types of friendship, ranging from forms in which friends simply provide one another with utility or pleasure to a form in which friends seek out not just the pleasures or benefits of association but far more: one another's virtues and excellence (their "real possibilities" some might say today).[11] The type of friendship from which we should consider learning is therefore not the friendship of long affection and intimacy, but the friendship of mutual concern, of care and respect for the other's practice of citizenship, their full participation in the political world. This is the friendship of appreciation of the hopes and political possibilities of the other, the friendship recognizing, too, the vulnerabilities of those hopes and possibilities.[12]

But neither friends nor stories promise—much less provide us with—decision rules for all situations. We get no gimmicks, no key to the inner workings of history, no all-purpose techniques for all cases. Instead we seem to get detail, messiness, and particulars.

That messiness of practice stories, defying our expectations, is an important part of their power. To some degree, the messiness is the message.[13] But saying that is far too simple. That messiness is important because it teaches us that before problems are solved, they have to be constructed, formulated in the first place. The rationality of problem solving, and the rationality of decision making too, depend on the prior practical rationality of attending to what "the problem" really is: the prior practical rationality of resisting the rush to interpretation, of very carefully listening to or telling the practice stories that give us the details that matter, the facts and values, the political and practical material with which we have to work.[14] If we get the story wrong, the many techniques we know may very well not help us much at all.

Consider finally now a skeptical challenge: Is all this about practice stories, and by extension all this about practical arguments, just about words?

It is certainly not just about words, for we have explored here what we do practically with words as we work together. We always face the danger, in studying ordinary work, that we will listen to what is said and hear words, not power; words, not judgment; words, not inclusion and exclusion; "mere words," and not problem framing and formulation, not strategies of practice. An Italian friend and colleague put this worry beautifully recently, when he wrote to a mutual friend and colleague, "No doubt, it is important to understand how [planners] behave in municipal offices. This is important for the sociology of organization and bureaucracy, for the analysis of policy, etc. and also for understanding a portion of planning implementation in practice. But for planning theory and practice it is less important than an apple [was] for Mr. Newton; in the end the apple is a metaphor, while what [Planner] Brown says in his office—it is just what he says in his office!"[15]

But what Planner (or policy analyst) Brown says in the office is not just what he or she says, though it is that too. What Brown says also embodies and enacts the play of power, the selective focusing of attention, the expression of self, the presumptions of "us and them" and the creation of reputations, the shaping of expectations of what is and is not possible, the production of (more or less) politically rational strategies of action, the shaping of others' participation, and much more. What Planner or Architect Brown says involves power and strategy as much as it involves words.

So our ears hear sounds. A tape recorder records what is said. Children might identify the words. But the challenge we face, as planners and policy analysts more broadly, is to do more: to listen carefully to the practice stories we hear and to understand who is attempting what, why, and how, in what situation and what really matters, what is valuable, in all that. That challenge is not just about words, but about our abilities to go on, our real opportunities and our actions, our own practice, what we really can, and what we should, do now.

2

Rationality, Emotional Sensitivity, and Moral Vision in Daily Planning Practice

We learn from friends as well as from scientists, from the accounts of thoughtful practitioners as well as from systematic studies. We learn from historical research as well as from philosophical argument and social science. We can learn from qualitative accounts and from quantitative analyses too, but we seem to forget all this very quickly when we try to understand the rationality of public decision making and public learning.

In planning and policy work, we plainly *do* learn from astute accounts of particulars, from the dramas of the moral challenges and conflicts of others. Watching and listening closely, we are impressed by some people and dismayed by others, but we often learn from both. Frustrated by some, we nevertheless pick up tips from others: we see new ways of going on, ways of handling pressure, ways of presenting information, ways of being careful and persistent.

Paying attention to those around us, we learn about character and our own possibilities at the same time. We can come to see more clearly not so much the brute facts as what seems to matter: what we take to be significant or at stake in the case at hand. To put it simply, we are likely to learn far more in practice from stories than from scientific experiments (Schön 1983).

Listening carefully to practice stories every day, we work to formulate our problems, to empathize with or understand others we hardly know, to deliberate and consider our own responsibilities and interests too (Hoch 1994, Krieger 1981, Mandelbaum 1987, Fischler 1995, Marris 1990). Yet often we reduce "acting rationally" to "choosing well," and so we neglect the challenging practical rationality involved in listening to practice stories astutely and in telling them carefully as well.

The very richness of stories that threatens their generalizability enables them to be so revealing: to show as well as to explain, to connect as well as to predict, to frame in a new light as well as to put an argument in context. Imagine, for example, that we want help before walking into a difficult meeting with a developer's lawyer and architect. Forced to choose, we are more likely to want the particular story of their previous negotiations with our department than we are to want a more general study of architect versus lawyer influence in developer-planner negotiations.[1]

But if stories are so often messy, detailed, particularistic, and unique, how can they help us to learn in practice? They can help precisely when messiness and detail, particulars and uniqueness matter: in an extraordinary number of cases, whenever real individuals must be treated not as stereotypes but as the specific people they are. The very messiness of thickly described practice stories has its own lesson to teach: before problems are solved, they must be constructed. Before we can consider options and choices, we must have a decent sense of what is at stake, of who and what is involved, to whom and to what we need to pay attention.[2]

In addition, because practice stories can convey the emotional demands of work in an ambiguous, politicized world, they can enrich our emotional awareness and responsiveness. Articulated in practice stories, emotions may teach us as well as move us. In empathizing with another's fear, for example, our emotional responsiveness may help us to see the world more clearly.[3] We may know abstractly that a developer fears the false assurances of permitting officials, that a neighbor to a site fears the "good intentions" of city hall, but we really know quite little here unless we know emotionally what it might be like to feel such fear in the particular circumstances of this developer and this neighbor. We must try to recognize difference and listen carefully, presuming neither that differences of experience, class, gender, or race, for example, must be unbridgeable and mutually incomprehensible nor that some perfect intersubjectivity will ensure equally perfect understanding.

When planners are distrusted and perceived as threatening and aloof, their failure to respond sensitively to those emotions will look not like professionalism but like callous blindness—if not a willful disregard for the well-being of others. In general, because we can see how those who are emotionally obtuse are blind to what those around them care about, we can see how emotional *sensitivity* works as a source of knowledge and recognition—and, as Nussbaum argues (1990a:75–82), as a mode of moral vision too (Coles 1989, Murdoch 1970, Kirp et al. 1989).

These claims suggest the agenda of this chapter which has four main sections. The first examines a practice story told by a planner, Harry, in a staff meeting of a small city's planning department. Even a quick look at Harry's story reveals it to be far from a simple tale about a multiuse project; it has a good deal to teach us.

The second section sets out the underappreciated aspects of practice stories: the ways they express and render political judgments, the ways their telling and listening require an astute practical rationality and responsiveness to the cases at hand. The third section examines another kind of practice story, a brief reflection of an environmental planner, Wayne, about mistakes he had made as a manager in his planning office. A brief but close look at Wayne's story shows that stories do far more than describe events. To think of "telling stories" as primarily describing events, we shall see, would be as limited a view as thinking of "playing music" as primarily making noise.

The fourth section turns in more detail to a story taken from a profile of a community development planner, Allan Isbitz, who recounts a deceptively simple vignette involving racism, risk taking, and astute political judgment. Allan's story suggests both the challenges of community development practice and the ways we all may learn from profiles of astute and perceptive planners. The conclusion summarizes the political, moral, emotional, and deliberative work to be done well (or ineptly) as planners tell and listen to diverse stories every day.

The Extraordinary Richness of Ordinary Stories in Planning Practice

In a midmorning staff meeting, called every two weeks to discuss work in progress, the planning director, Tom, says to his half-dozen planners, "We've got a problem on the Northside. Can someone fill us in, since I have to go at noon?"

Harry, his assistant director, responds:

The problems on the Northside? Let me try. We've got four users here: the park, the Housing Association, the Nature Center, and the Children's Center. And you've got three different architects working on the project, and no overall set of assumptions for what the constraints are for what they're designing. And then you've got Charles [the city attorney] negotiating various land sales, also without regard to any kind of overall design layout, particularly for access and circulation, and that seems to be the crux of the problem right now.

There was a very early site plan put together that suggested a portion of Benjamin Street could be used for parking, and that there'd also be bus access to the front of the Nature Center site and the Children's Center site along Benjamin, and that that bus would probably get there through a new road that would be cut between the park portion and housing portion of the site.

In our negotiations with the Housing Association—that is to say, Charles and myself on behalf of the Intergovernmental Relations Committee—we really tried hard to structure a deal that's not going to break their backs, and one of the things we said to them is that, "Look, your earlier proposal suggested that the city was going to build the street, and that's just not in the cards. The city is not going to build a street; you guys are going to have to build it. And it seems to us that one way to sort of build something that gives you the frontage you need for zoning—because they've got this ridiculous requirement of having to subdivide this site into fourteen individual lots—would be to build a small, private street that would give you frontage and would only be used for the people whose houses front that street.

Here another planner, Lynn, asks: "What's ridiculous . . . what's the ridiculous requirement about having to divide this into fourteen lots?" and the staff go on to discuss the project. They ask about the zoning requirements, the actors involved and their commitments, the encompassing political pressures, and so on.

The assistant director here has told a quite ordinary story in an ordinary staff meeting. The staff take his account of what is happening on the Northside as reasonably ordinary too. There are problems, surprises, fights, negotiations, personal frictions—the usual uncertainties. So the planners listen for the relevant details; if they can make no helpful suggestions, at least they can better understand a coworker's problems and perhaps his best judgment of what is to be done now.

But this story, like many other stories told in practice, is really quite extraordinary. The story not only presents but constructs a problem; it not only identifies actors ("four users here"), but it characterizes their collective irrationality: "And you've got three different architects working on the project, and no overall set of assumptions for . . . what they're designing."

There's much more to this story too: attention to particular actors (Charles, "negotiating various land sales") in the context of more general planning concerns (Charles's negotiating "without regard to any kind of overall design layout"). And more: a mention of relevant institutional and normative history (the "very early site plan"), before the account of Harry's own efforts in the recent negotiations: "We really tried hard to structure a deal that's not going to break their backs." The story begins to put the project in its current institutional context too: "They've got this ridiculous zoning requirement of having to subdivide this site into fourteen individual lots"; and "The city is not going to build a street; you guys are going to have to build it."

The paragraphs quoted above are only one small part of an evolving story of "the problems on the Northside."[4] But the richness of even this segment of the story teaches several lessons.

What appears initially as a report of recent events quickly becomes a complex set of future issues the staff might explore. Harry is not just providing facts; he is presenting the facts that he takes to matter, facts he takes to be significant or even worrisome. What we call ordinarily "setting out the issues" is really quite extraordinary, ethically speaking, for it means making judgments about what is important, valuable enough to need further attention and study. This deliberative aspect of storytelling and listening calls for what academics refer to as deconstruction and reconstruction on the job, practically and ethically. What, for example, is Harry saying about Charles: Is he a loose cannon, acting willfully "without regard to any kind of overall design layout," or is he being driven by the mayor? What about those "ridiculous" zoning requirements? Just what does Harry mean, and what is really important here (Manin 1987, Reich 1988, Majone 1989, Vickers 1984)?

Such practice stories portray not only issues but a practical world of action and interests, of settings and histories, of strategies and counterstrategies. Notice that Harry cannot define the context of the Northside project once and for all, because that context is always changing. That context changes as the local economy fluctuates, as the city council becomes more or less attentive, as state legislation evolves, and it changes practically with every move made in the current negotiations, with every promise or threat.

Harry's account suggests not only several of the actors involved, but the precariousness of their relationships. The architects are working without a shared sense of constraints; the Housing Association had been expecting the city to build a new street, but Harry and Charles have told them, on "behalf of the Intergovernmental Relations Committee," "that's just not in the cards." We learn too about contingencies: the housing project's success will depend on a "ridiculous requirement of having to subdivide this site into fourteen individual lots."

Harry portrays in his story a practical and moral complexity of differing expectations, legal obligations, governmental commitments, historical precedent, and design suggestions as well. The story helps his planning colleagues to understand better not only the Northside but themselves: to appreciate how Harry and, by extension, each of them as the city's planners are morally entangled, morally obliged to listen to the architects, to respect but perhaps attempt to rein in Charles's negotiations, to confront the requirements of the zoning law, to recognize but improve the early site plan, to negotiate successfully with the Housing Association. These entanglements are both moral and practical requirements, each of which can be met well or poorly, each of which may affect the success of the ultimate project. Perhaps most important, planners' practice stories can convey the moral complexity of issues and the practical moral entanglements of planners in ways that more abstract accounts of planning cannot. When sensitively told, these practice stories can teach listeners what it may be like to face, to "be finely aware and richly responsible to," the issues they portray (Nussbaum 1990a; this volume, chap. 5).

Beyond Description to Practical Rationality: Stories Render Judgments

We could still insist on looking at planners' practice stories as descriptive accounts. Seen this way, the stories would be pictures—snapshots and approximations of events and behaviors. We would often discount these stories a good deal, knowing how much they inevitably leave out.

But we have a better alternative. We can look at these stories no longer as primarily descriptive but as prescriptive: telling us what is important, what matters, to what we should pay attention, what we need to worry about, what is really at stake if we fail to act. When a planner asks a colleague at work, "What happened at the meeting last night?" she is far more likely to be asking, "What happened last night that I might need to know about?" than "What really happened, step by step?" Seen in this light, these stories set agendas, shape senses of relevance, contribute to priority setting, construct problems, and shape action.

No longer tales told simply to entertain or describe, these stories now appear ethically selective through and through. If we listen closely, not only to the portrayals of fact in planners' stories but to their claims of value and significance, we discover an infrastructure of ethics, an ethical substructure of practice, a finely woven tapestry of value being woven sentence by sentence, each sentence not simply adding, description by description, to a picture of the world, but adding care by care to a sensitivity to the practical world, to an attentiveness to and a prudent appreciation of that world. We learn from skillful (and perhaps inept) performance as well as from verified (or refuted) propositions.

Iris Murdoch (1970:37) makes the more general point powerfully: "If we consider what the work of attention is like, how continuously it goes on, and how imperceptibly it builds up structures of value round about us, we shall not be surprised that at crucial moments of choice most of the business of choosing is already over. This does not imply that we are not free, certainly not. But it implies that the exercise of our freedom is a small piecemeal business which goes on all the time and not a grandiose leaping about unimpeded at important moments. The moral life, on this view, is something that goes on continually, not something that is switched off in between the occurrence of explicit moral choices. What happens in between such choices is indeed what is crucial." Again: before the rationality of choice comes the prior practical rationality of careful attention, critical listening, setting out issues, and exploring working relationships as pragmatic aspects of problem construction.

Planners build up such structures of value in their stories in institutionally and ideologically staged ways. Certain values seem excluded, to arise only rarely, if at all; others are "impractical" or "not serious" or "out in left field" given the staff's reading of those with whom they work.[5] Planners' stories inevitably express relations of power, reproducing those relations in politically diverse ways (Tett and Wolfe 1991, Healey 1993a, Campbell and Fainstein 1996, Flyvbjerg 1998).

So we can recognize in planners' practice stories the ongoing rendering of political judgments in institutional settings in which resources can be few and ambiguities many, in which authority can be minimal but vulnerability ample. Planners' practice stories reflect and render political judgments because the planners have not just issues to face but relationships to sustain: with politicians, citizens' groups, official boards, legal staff of city and developer alike, architects—adversaries and supporters too. In a conflictual political world, the failure of these relationships can stop the planners' efforts cold. The city council may withdraw support. The head of the housing authority may withdraw cooperation. The supposedly friendly planning board may stall. The citizens' group may distrust the staff's promises. And so on.

Sorting through an infinity of fact, facing ambiguous desires, interests, and mandates, the planners must search for an account of the issues and actors so they can fashion working agreements with others on this design or that strategy, on that project mix or this schedule. They are neither just describing facts nor simply prescribing values; they are searching for possibilities of agreement and consent, for others' support, for a solution that will make sense to others as well as themselves (Benhabib 1992).

Tore Sager (1994a) quotes Hannah Arendt on the centrality of judgment to astute deliberative practice: "The power of judgment rests on a potential agreement with others, and the thinking process which is active in judging something is not, like the thought process of pure reasoning, a dialogue between me and myself, but finds itself always and primarily, even if I am quite alone in making up my mind, in an anticipated communication with others with whom I know I must finally come to some agreement. From this potential agreement judgment derives its specific validity." Anthony Kronman (1987:869) puts the other-regarding aspect of political judgment in more pragmatic terms: "Anyone wishing to be effective in debate will have an interest in becoming the sort of person whose opinions are respected, . . . a person of good judgment."

But in a staff meeting, how do the planners' stories reflect their "anticipated communication with others with whom they know they must finally come to agreement"? The planners' time is short; their agendas are full. Politically vulnerable in the face of power, the planners know that they will have to make a case to citizens, developers, and politicians alike. Thinking through any project's problems, planners are continually under pressure to search not just for what is good in some abstract sense but to find what is good in the political sense of potentially gaining the approval of others.

Harry's story, for example, suggests his search for strategies to move ahead: he recounts his proposal for dealing with the "ridiculous" zoning requirement at the Benjamin Street site. "We tried really hard," Harry says of his and Charles's efforts.

Harry does not simply complain about the zoning and portray himself as a victim; he describes the zoning requirement, his effort to respond, and the practical suggestion he has made to move ahead—a suggestion he brings to the staff meeting for discussion and refinement. So, too, the stories planners tell at work can reveal political judgments about opportunities and constraints, about more and less responsible efforts, about more or less supposedly legitimate mandates, about relevant history to be respected and learned, relevant concerns, interests, and commitments to be honored.

These processes of judgment are remarkably complex—certainly as complex as the work of description!—and they are as remarkably understudied. We need to learn more about both the kinds of judgments planners must routinely make and the experience, education, and sensitivity planners require to make such practical judgments well.

An Environmental Planner's Confession

The stories that planners tell about their own work suggest the rich variety of judgments they must make in practice. Consider now two excerpts from a series of profiles of planners (Forester and Chu 1990; Forester and Kreiswirth 1993a, 1993b, 1993c, 1993d). The first suggests how much more than description a planner's story can accomplish. The second suggests ways that planners' practice stories can teach us about the riskiness and opportunities of practice, the moral imagination required, and the kind of practical rationality arguably necessary as well.

To begin, consider a passage from a profile of Wayne, an environmental planner (Forester and Chu 1990:197). Let us see if Wayne's story, superficially about management style, is simply a description of facts, or much more.

Speaking of his experience of managing a planning team, Wayne tells us:

> That stuff I had no clue about. So you'd be real rigid about your management style when it wasn't appropriate, and people resent that. So you learn later to be rigid when you need to be rigid and lax when it's okay; and that way everybody gets a balance. They don't feel so bad about working hard later, because it's been lax other times; and you're getting the most out of it. You're using it when it's appropriate. That's an important lesson.

Perhaps this little story of learning on the job could be considered a description, but in telling this tale, Wayne does much more than recount facts. We see Wayne self-report, *admit and confess* his own *ignorance:* "That stuff I had no clue about." He *identifies* his own, and by extension others' *mistakes:* "So you'd be real rigid . . . when it wasn't appropriate." He *predicts* instrumental, psychological, and political *costs* of a rigid management style: "and people resent that." He also seems to *acknowledge* a substantive *harm* of causing staff resentment, independent of its instrumental hindrances.

Yet he *recognizes* hope and *envisions* possibility too: "So you learn." He *makes the judgment* that contingency counts, and he suggests what can be learned: "to be rigid when you need to be rigid and lax when it's okay." He *empathizes* with his staff about their working conditions: "and that way everybody gets a balance." He *appreciates* staff morale: "They don't feel so bad about working hard later," and he *honors* a norm he implies his staff shares, a norm of reciprocity and reasonableness: "because it's been lax other times." Nonetheless, he *justifies* "laxity" strategically: "you're getting the most out of it"—and he *legitimates* his own managerial action: "You're using it when it's appropriate." Then he *sums up* and *evaluates his* own short story: "That's an important lesson."

Wayne's story expresses moral imagination as much as empirical description because, like many other planners' stories, it is not just about facts, but about facts that matter, that are significant to planning practice and citizens' lives. Wayne's work of description is also one of confession, empathy, recognizing possible harm, respecting social norms like reciprocity, and morally imagining possibilities and opportunities.

The point here is not to show how complex a simple paragraph can be, but rather to debunk the idea that planners' practice stories are primarily descriptions, pictures of events, mirrors of some unambiguous reality. Telling their stories, planners act in very concrete, very specific ways, making particular practical judgments for which they are often held responsible—recognizing or missing important issues—by those with whom they work.

A Community Development Planner on Racism, Particularity, and Hope

Turn finally to an excerpt from a profile of Allan Isbitz, a planner hired as executive director of a settlement house seving poor white Appalachian and African American communities (Forester and Chu 1990:20–24). Allan tells us:

The first thing I did was advertise for a home buyer counselor. I remember saying, "If we're going to have an equity program here, we're going to have to buy these houses, rehab them, and sell them." And I needed somebody to advise residents who had never before dealt with banks, on how to deal with banks. How to establish credit. That was the first thing I started with.

I was also aware of the split between the black community and the Appalachian community. I knew that in order for this arrangement to work, they had to work together and present a unified neighborhood front for the economically poor in that neighborhood. Any differences between groups would be played off by those who didn't want them to succeed in the first place.

Allan had moved from a city with a strong history of nonprofit community development work to what seemed to be another world:

I found myself in the bastion of the private sector. People in this city didn't really respect government, except insofar as it supported the private enterprise movement. The whole rhetoric was different. The city only had experience with a couple of non-profits that, by and large, had failed. And the only reason they kept pouring money into those institutions was because of straight political deals.

The *only* person I found for this home buyer counseling role was this woman who responded to the ad I put in the paper. She had grown up as a welfare mother herself, broke out of it and got a home in the private rental market. She was also on the staff of the Housing Authority counselling public-housing tenants on how to get their financial act together when they fell behind in their rent. . . .

She could understand what I was trying to do to get to the poor Appalachian and black communities, who had no experience in dealing with banks. I was saying, "What we want to do is put together a credit program for the families, give them six months of working with a home buyer counselor before they even set foot in a bank, have the home buyer counselor walk into the bank with the family, with all the documents laid out that the banks are used to, and present a credit-worthy risk so that the banks will finance these individuals." She could understand that. I knew there was a lot of training I had to do, but she could understand the program. She just *happened* to be black.

The advice I got, though, was, "The Appalachian poor will *never* work with a black. You're taking a risk with the project."

Well, I didn't have the luxury of asking questions. She was the only person I found who I thought could do the job. The only thing I asked her during the job interview was: Did she mind working in an environment in which a lot of people would be prejudiced against her? I asked her point-blank: How comfortable did she feel in dealing with Appalachian whites? I didn't have to ask that question twice—she knew exactly what I meant. She said she thought she could do it.

I left it up to good faith after that. In any environment, there are prejudices that people face. But when you start to work one-to-one, when you start really to help people who need help, and do it in a way that is not heavy-handed and maximizes their options for succeeding, the effort is appreciated. I was banking on the fact that the poor communities would appreciate what we were trying to do, even if they didn't in the beginning. A lot of people in fact wouldn't work with us in the beginning. But ultimately, they did. She was good, sensitive, and not overly aggressive. In fact, she was very quiet. In her own quiet way she plugged along, helped everybody over the two years, and became a very trusted member of the Appalachian community.

How can we learn from these few paragraphs? Certainly we do not learn that the success of the story is generalizable. Neither do we learn that the case is unique. What then?

We learn, first perhaps, about the crucial importance of particular people in particular places and times. Allan's story leaves us with no doubt that his program might have failed, that his good intentions might have gone disastrously wrong, that one other counselor might have done far less well.

We learn about the immediate and pervasive threat of racism to the planning process. Allan had been told of his chances: "The Appalachian poor will *never* work with a black." That prediction was no neutral statement of likely fact; it was a direct warning: "You're taking a risk with the project." The pressure—if not the risk—here, and no doubt in other cases, was inescapable, Allan recounts: "I didn't have the luxury of asking questions," of exploring alternatives, of taking more time.

But Allan did not take that risk blindly, exposing not only the project but the prospective black counselor to the racism of the poor Appalachian community in which she would have to work. He spoke directly, if not singlemindedly, to the issue to recognize and assess *with her* the threat of racism from the beginning: "I asked her point-blank: How comfortable did she feel in dealing with Appalachian whites?" This was, he implies, no easy question to ask. We would hardly say ordinarily, for example, "I asked her point-blank if it was raining."

Allan suggests too their developing collaboration: "I didn't have to ask that question twice—she knew exactly what I meant." Here we see less a meeting of minds between a white male planner and a black female home buying counselor than we see a glimmer of their joint recognition of a common threat.

Allan thought she could do the job. He acknowledged the threat of racism, asked if she thought she could still work in the face of it, and when she said she thought she could, he tells us, he "left it up to good faith after that." This appeal to "good faith" was pragmatic, not theological: a faith in a sensitive practice and a sensitive, experienced practitioner, and *not*, significantly, a faith in an abstract solution, a way to handle racism in general, a recipe for community development, a "one, two, three" method of community work.

Allan tells us instead of good work in the face of the perversity of racism. He tells us too of the kind of rationality required here: a kind of practical rationality that does not promise grand strategy but responds to particular need, that does not so much fix clearly on an end and choose a means to it as it responds without creating new problems, in a way, Allan tells us, "that is not heavy-handed."

But there's a good deal more going on here. We learn too from the emotional quality of the story. We learn not only about the iffyness of practice, its contingencies and fragility, but about the planner's and counselor's experience of that iffyness, their experience of real vulnerability. Allan's efforts to hire a counselor for applicants to the home ownership program brought him an underwhelming pool of a single qualified counselor. This might have simplified Allan's choice, but it can hardly have given him any real sense of making a choice at all. So rather than seeing

Allan managing a program, we see him instead being managed by the constraints he faced and feeling vulnerable to the resulting dangers.

We learn about a kind of emotional particularity here too, for Allan's single qualified applicant knew about more than bureaucracy, housing, and banks. Allan's very first words about her suggest important parts of her story: "She had grown up as a welfare mother herself, broke out of it, and got a home in the private rental market." She knew personally the poverty of the communities Allan hoped to serve; she also "broke out of it," an expression suggesting not happenstance but a personal story of struggle. And those personal qualities, we come to learn, were crucial: "sensitivity" to the people she worked with, determination and persistence too ("she plugged along, helped everybody over the two years").

Particular sensitivities and luck matter as much as technical competence here. Speaking of a different case, Allan put it succinctly: "You find people all of a sudden. Well—you have to have your eye out. You have to know what you're missing and have an eye out for what you need." So luck matters, but so too does a care-full perception, so too does the work of search, knowing what you are missing and having an eye out for what you need. The judgment Allan had to make here—to hire this applicant or not—involved creating his own program's luck, enabling "the right person to be in the right place at the right time" (Pitkin 1984).

We see that Allan knew what he needed from his prospective counselor. He saw that she had worked for the Housing Authority and understood his program, but he knew too that both she and his program would be vulnerable to racism, so he looked not only for technical qualifications but for what we can call "moral qualifications": the qualities of real responsiveness under pressure that we call good judgment, wisdom, or, more colloquially, street smarts. Allan knew to search *not* for idealistic rule following, not for a moral saint, but for tested virtues of character: an open-eyed recognition of the threat of racism and a considered response in the face of it. "How comfortable did she feel in dealing with Appalachian whites? I didn't have to ask that question twice. . . . She said she thought she could do it."

Allan tells us a great deal here about himself too. He did not feel the need to ask his question twice; he felt confident that she recognized what he was talking about; he accepted her response, her judgment that "she could do it."

Allan has another lesson to teach us in his emphatic, "She knew exactly what I meant." Literally minded theorists might object here: How could Allan know this? How could he know what she meant, at his distance from her African American, gendered experience? But this objection would miss the point, besides presuming the kind of full knowledge it seeks to criticize, for Allan is not making philosophical claims about sameness of meaning or experience. He is making a practical claim about their interaction, and he does it with hindsight, claiming that given the situation, his words and her response, and their subsequent behavior, she knew exactly what he meant regarding the practical purposes they both recognized.

We learn here about two people's courage and determination, and we know that while one faced the loss of his program's success, the other faced the potential loss of life and limb. We can hardly understand this story, it seems, if we fail to understand the relevant emotions involved here: his directness, her courage, their collaborative "understanding," his confidence in her abilities, her sensitivity, his conviction "that the poor communities would appreciate what we were trying to do," her one-to-one help of people in need, his willingness to move ahead in the face of "a lot of people" who "wouldn't work with us in the beginning."

These emotions are far from irrelevant here and far from irrational. Indeed, only if we recognize these emotions can we even begin to make sense of the actual practice in this case. Allan took risks, we learn, not blindly or callously, but deliberately, if also with good faith as well as trepidation—combined no doubt with hope and fear on the counselor's part too.

Similarly, we can hardly doubt that the counselor's success grew not just from her competent provision of information about bank practices but from her particular emotional qualities too: her persistence, her sensitivity, and her "not overly" aggressive directness—emotional qualities of attentiveness and responsiveness in practice without which she might well have failed. So Allan's story suggests the importance of intellect and emotion, technical competence and affective responsiveness, as complementary if not interpenetrating aspects of practice, required of practitioners whatever their class, race, and gender. An important empirical question in any given case is whether, how, and when particular practitioners of varying class, racial, and gender backgrounds might embody differing sensitivities and abilities, given the requirements of working in that case—with immigrant Asian women, with Eastern European gypsies, with African American Baptists, and so on.

These emotional qualities of practice that Allan sketches for us seem not incidental but crucial. Hardly irrational, these emotional qualities of a black woman's rich responsiveness to poor whites seem to have enabled her to have overcome racist suspicions, to have succeeded programmatically, and to have enabled her against the apparent odds to have become "a very trusted member of the Appalachian community."

Allan's story presents a tale of emotional struggle and judgment interwoven with a program's development. To neglect the emotional character of the story would make it less subjective—making us less able to empathize with and understand the characters involved—and *less objective* too, telling us less of what really mattered in the case at hand, less of the actual qualities that enabled Allan and the counselor to act effectively.

Allan's story suggests more too: that emotional sensitivity and responsiveness work together, along with technical knowledge, as a mode of practical response and that such sensitivity and responsiveness are not simply incidental accompaniments of cognition, of "thinking." *This no more means that any emotion any time will do, any more than it means that any fact is relevant.* It means instead that in a world of difference, emotional sensitivity can be a form of moral vision, of moral attentiveness to others (Nussbaum 1990a:75–82, Rorty 1988). It suggests too that planners lacking emotional range, emotional maturity, and capacity will likely miss a good deal of what lies before them, and they are likely to fail as a result.

As Kronman (1987:858) argues, "To deliberate well—which requires both sympathy and detachment—one must . . . be able not only to think clearly but to feel in certain ways as well. The person who shows good judgment in deliberation will thus be marked as much by his affective dispositions as by his intellectual powers, and *he will know more than others do because he feels what they cannot*" (emphasis added). So Allan speaks of the importance, for example, of not being "heavy-handed" or "overly aggressive," and of his faith that their efforts would be "appreciated," recognized as valuable and not self-serving, as promising and not as a con.

Allan has learned too. He took a substantial risk, and his retrospective story reflects what he has learned, as well as what he can now reconstruct briefly about what really did happen. He has learned, for example, about racism and practical work. He did not seem to take lightly the warning that he was risking the project, yet he took a risk, and certainly exposed the counselor to greater personal risk, without full information, without a well-developed means to his desired ends. But he was not flying blind. He tells us what his bets were, what his working theory was: "I was banking on the fact that the poor communities would appreciate what we were trying to do, even if they didn't in the beginning." In hindsight, those bets and that working theory seem to have been vindicated, we learn, as Allan learned.

We should explore the kind of judgment, the kind of rationality, that Allan employed here. He did not simply satisfice, lowering his expectations to find a satisfactory outcome, nor did he muddle through by building on piecemeal, incremental agreements with stakeholders to move ahead. On the contrary, he tells us: "A lot of people in fact wouldn't work with us in the beginning." So Allan did not work by consensus building either; quite the contrary, in hiring a black woman to work in part with a poor white community, he was working directly against the practical advice of others.

Allan seems to have recognized not just one but several very general, and very important, facts, encompassing histories, that were operating in the situation he faced: the histories of racism, first, and poverty, second. The poverty of the African American and Appalachian communities had motivated his project in the first place; the racism of the Appalachian community, he tells us, threatened the same project. Yet Allan wasn't stuck here. Why not?

In the face of warnings that he was endangering the project, in the face of "a lot of people who wouldn't work with" him, what made it possible for Allan to move forward and not bail out, not wait for another day? His story suggests that he saw more than the facts of poverty and racism: he recognized the human capacities of others "to appreciate" the effort "to help people who need help" in a way that is "not heavy-handed." As Nussbaum (1990a) suggests, Allan's practical judgment reflected a dialogue of antecedent principle and new vision, a recognition of both general normative principles and the particular possibilities of this case.

Had Allan taken a vote in the affected community initially, he might have closed up shop. Had he taken a vote later, he might well have been supported. In the language of rational choice theorists, Allan bet that the community's initial preferences—refusing to work with a black woman housing counselor—would change, that the general threat of racism could be overcome by "sensitive," "one-to-one" work *in this case* (March 1988). But this case is what matters, if we are to understand the possibilities of case-specific practical judgment, the connections between thinking and acting, learning and doing, perception and action in actual practice.

But did Allan bet that the community's preferences would change? He tells us that after the prospective counselor told him she thought she could work with the white community, he "left it up to good faith after that." With the benefit of hindsight, and somewhat glibly now, he tells us that "in any environment, there are prejudices that people face." Had he been that cavalier originally, he never would have asked the counselor "point-blank: How comfortable did she feel in dealing with Appalachian whites?"

If Allan can now see that these prejudices were not determinant, he hardly knew that with any certainty in the beginning. If Allan "bet" on the counselor's success, he did it not by throwing the project to the winds of chance but as a participant, expecting to do "a lot of training," expecting that the particular person, her particular style, and the particular community's needs could mesh and be helped to fit together successfully. Allan's story suggests no weighing of commensurable costs and benefits. It suggests instead a careful work of practical rationality, a careful consideration of general facts, including racism and poverty, that would influence the actions of particular people with particular sensitivities and histories and which would, in the specific case, be influenced in turn. This rationality, of Allan's and the counselor's, is interpretive and practically pitched, responsive to the ambiguity and incommensurability of real situations, and this rationality is conveyed powerfully by practice stories, whose messiness and moral entanglements characterize the work that practically rational actors must do all the time.

Conclusion

This chapter has explored the practical, political, and ethical character of the stories that planners routinely tell—at work and about their work too. The argument in a nutshell has been this: by telling practical stories, for better or worse, planners bring to bear moral imagination and shape the moral imaginations of others too.

Telling practice stories, planners render practical and political judgments not in their minds but in their deeds, as they come to be responsible for reporting or failing to report events, for recommending or failing to see options, for identifying threats or opportunities or failing to, and so on. In so doing, they search for value, set agendas, characterize others, define constraints and possibilities too.

Telling stories at work, planners not only describe events, but they may explain what has happened, warn of dangers, and identify "benefits"; they report relevant details and search for others' meaning ("I think what he meant was . . ."); they confess mistakes, justify recommendations, prepare others, and do far, far more. Doing moral work, planners' stories characterize others by giving them status or ascribing stigma. Doing political work, planners' stories organize attention, including some concerns and excluding others; invoking or challenging supposedly legitimate norms, pointing to support or opposition, future problems and possibilities. Doing emotional work, planners' stories allow listeners to empathize at times, to see issues more clearly because they feel more sensitively too.

Planners' stories do deliberative work as well. Always told in constrained circumstances, these stories reflect their tellers' ongoing search for value, for what matters, for what is relevant here, what is significant. Planners' stories are thus ethical not because they reflect right or wrong decisions, but because they reflect appreciation of what matters in the case at hand or blindness to it, because they reflect a responsive awareness to what is at stake or, alternatively, insensitivity, because they reflect a responsible appropriation of norms and precedent or a callous disregard of them.

If we wish to understand the everyday politics, ethics, and rationality of planning, we need to appreciate the complex practicality of the stories that planners must not only listen to but also tell in turn. If we think of stories simply as descriptive tales, we will miss most of their richness and significance, the challenges and opportunities, the care-full work and the serious blindness of actual planning practice. If we are to interpret the world of planning in order to change it, we should pay attention not just to the partiality of citizens' and planners' stories, but also to the real moral and political work those stories do. Only by listening critically and responding sensitively, beyond mere words and good intentions, can the good judgment of deliberative practice reconcile day-to-day pragmatism with moral-political vision.

II

Consensus Building and Mutual Recognition Create Deliberative Opportunities

3

Challenges of Mediation and Deliberation in the Design Professions

Because planners and designers work in the midst of many interested parties, they inevitably work in the face of conflict.[1] To do that, they need to improvise creatively and proactively; they must often act as both negotiators seeking desirable ends and mediators managing the conflictual planning or design process itself (Forester 1989:chap. 6, Susskind and Cruickshank 1987, Rivkin 1977). This means, though, that planners and designers can do far more than chase after compromises: they can promote effective processes of public learning, practical and innovative instances of public deliberation, even consensus building in many parts of the larger planning process (Hoch 1994, Healey 1997, Innes 1996, Menkel-Meadows 1995).

To serve both practical and theoretical purposes, this chapter examines two practitioners' accounts of their work to foster deliberative planning processes. Practically, this analysis seeks to describe more clearly the complexity of planners' and designers' roles, the diverse ways these practitioners can promote deliberative public learning, and the many forms of their persistent, imaginative questioning of design options. Theoretically this analysis seeks to show more clearly how planners and designers can encourage public learning about social significance and value as well as about positive fact, about historical identity and difference as well as about shared common ground (Abers 1998, Abrams 1991, Benhabib 1996, Sandercock 1995, Susskind, McKearnan, and Thomas-Larmer 1999).

The challenges of deliberation in the design professions are, most simply, the challenges of learning about what to do. That, not so simply, means learning about what we *should want* in a specific case as well as learning about how to get it, learning about appropriate ends as well as about effective means. Such learning then embraces not only facts and functions, data and capacities, but what is important or valuable in a case, what is to be honored or protected, encouraged or developed. We can see such deliberative work in many settings: in design reviews, neighborhood meetings, meetings with another agency's staff, even in staff meetings where planners consider what they should do and how they should act (Hoch 1994).

In deliberative work, citizens integrate the worlds of "is" and "ought," of "science" and "ethics," as they learn how to get something done and what ought to be done in new and unique cases too (Gutmann and Thompson 1996, Nussbaum 1990a, Wiggins 1978). Connecting governance and dispute resolution, politics and ethics, deliberative practice involves the most intellectually intriguing issues in the planning and design fields: how we can learn not only about technique but about value; how we can change our minds about what is important, change our understanding and appreciation of what matters, and, more, change our practical sense about what we can do together too.

Furthermore, as the next chapter will show, in community planning deliberations we can also learn more about who we already are—as inheritors of a distinctive history, for example, or as contemporary neighbors coming to terms with one another. We can learn about new relationships we can develop, and about who we may yet come to be together as we reshape our shared public world. All of these issues of "is" and "ought," and historical, present, and future identity, are at stake, potentially explored or neglected, in many of our planning and design processes (Barber 1984, Reich 1988, Sandercock 1995, Schneekloth and Shibley 1995).

In the face of conflict, when distrust is high and confidence low, we can easily miss the subtle practical payoffs of deliberative discussions. Jane Mansbridge argues that even sophisticated accounts in political science often ignore the real probing and transformation of interests that occur in political processes. In the planning and design professions, too, we may lapse back into truisms of "compromise," "fundamental" differences, and "trade-offs," as we then fail to realize how parties can learn, how their wants, interests, preferences, and priorities can shift and evolve in planning and design deliberations. Mansbridge (1992:42) writes:

"Negotiation" denotes a mix of power and influence. In negotiation, the parties involved not only maneuver for advantageous positions, as they do in conflict; they also try to understand what the other really wants, in order, for example, to offer what may be a cheaper satisfaction of that want than what the other is demanding. The quest for understanding requires asking and listening, correctly interpreting the other's language and putting oneself in the other's place. It requires making suggestions that the other may not have thought of, and learning from both acceptance and refusal. When negotiators engage in this quest for understanding, they can use the understanding so gained to change the other's preferences. They can help others discover what they really want, creating new preferences that better reflect the other's needs or values. They can even help others develop new values. Successful negotiations in the real world rarely rely on mere jockeying for advantage in the conflict. Successful negotiators often find ways of meeting one another's real needs at less cost than seemed originally required.

We do ourselves a great disservice if we think about planning and design deliberations as mere "process," periods of potentially distracting and draining "talk," a necessary evil accompanying the "real work" of planning and design. Planners and designers shape not only physical spaces, but what Lynda Schneekloth and Robert Shibley (1995) call "dialogic spaces," deliberative or broadly argumentative spaces as well. Those dialogic or deliberative spaces include the meetings (often the hundreds and hundreds of meetings) occasioned by the projects that planners and designers work on, and the related negotiations, discussions, project reviews, charettes, hearings, and review sessions that bring affected citizens, regulators, developers, and public officials face to face (Abers 1998, Innes 1996, Katz 1994, Riddick 1971, Susskind, McKearnan, and Thomas-Larmer 1999). Hardly mere process, these deliberations can reproduce public imagination or blindness, public hope or cynicism (Susskind and Field 1996). They might strengthen citizens' capacities to listen and engage one another, or they might instead degenerate and encourage the all-too-common intransigent public hearing postures of "decide-announce-defend": "First I'll decide what I want; then I'll announce it; and then I'll defend it!"

We should keep alive two related conceptions of public space: those of the physical Newtonian spaces of plazas, parks, and avenues in which streams of diverse citizens typically pass each other by, and the argumentative public spaces of municipal offices, conference rooms, legislative chambers, school auditoria, living rooms, and small town meeting places like the firehouse or church basements in which community members jointly envision, criticize, and refine the design and redesign, the preservation and development, of our cities and towns (Faludi 1996, Fischer and Forester 1993, Kemmis 1990, Nelessen 1994). If planners and architects focus only on physical objects, they will hardly appreciate how they can weaken or strengthen the deliberative public spaces through which they must do their work.

As we shall see, effective deliberation in planning requires attention to both the substantive issues at hand and the relationships that link the parties who care about those issues. Because planning professionals can create deliberative spaces, they must have the strength to listen to strongly held but conflicting views. Always seeking to learn as they go by asking questions, they must be able to distinguish deeper concerns from more superficial rhetoric, so they must be able to listen perceptively and come to see issues anew, and at times they must be able to teach as well. They must pay attention to product and process, to arguments and argumentation, and they must do this in the face of institutional rivalries, uncertainty, and conflicts about "what ought to be done." These planners and designers face the challenges of playing multiple roles simultaneously. At different times within the same process, they must bring the knowledge of experts; they must listen and encourage creative solutions as mediators; they must defend particular values as negotiators; they must structure processes of participation, discussion, invention, and decision making as organizers too.

To explore how planners and designers can improvise to play these multiple roles, we can now build on chapters 1 and 2 by considering two accounts of design professionals who have been working in the face of public conflict. These practice stories are excerpted from profiles of planners and designers from Israel (Forester, Fischler, and Shmueli 1997) and Norway (Forester 1994a). These accounts span the fields of architecture, archaeology, historic preservation, environmental planning, and land use control. They can teach us not only about issues of land use and urban design, but also about the challenges of the deliberations that planners and designers can foster and manage in their daily work (Hoch 1994, Healey 1997).

In each case, we can learn from practitioners who have reflected on their deliberative work, the threats to that work, and the mixed success of their responses. If we listen closely, and bring a theoretically informed ear to these accounts, we can also identify regions of planning and design practice in which that practice clearly leads, and can teach, theory. But even if we should not expect recipes here for "how to do it" any more than we would expect recipes for how to be a good friend, son or daughter, parent or teacher, we can still look forward to insights about the real challenges and opportunities that planners and designers face.

To begin, an architect-planner's detailed story of archaeology and architecture, tourism and economic development introduces the complexity of the issues. In project planning for the Old City of Akko, on the northern coast of Israel, multiple actors and concerns come into play. Several conceptions of public space interact. Multifaceted building processes overlap. Persistence and a sense of consensus building lead to promising results. Next, a deceptively simple example of a planning subcommittee shows both quite ordinary work and the potentially extraordinary roles that design professionals can play.

An Architect-Planner's Embrace of Ancient Stones and Modern Hopes

To begin to explore the deliberative aspects of planning and design practice, listen first to the story of Arie Rahamimoff, an architect-planner born in Bulgaria, trained in Israel and Finland, having taught in the United States (at Harvard and the University of New Mexico) and in Germany as well. Rahamimoff tells us about historic preservation and economic development, about tourism and social policy, about professional blinders and consensus building all at once. We can consider his story in his own words, but in several parts. The historic port city of Akko in Israel lies just north of and across the bay from Haifa, less than an hour's drive south of the Lebanese border, with Nazareth a half-hour's drive to the east. The walled Old City is a predominantly Arab section of the larger modern city of Akko that has developed around it.

Rahamimoff described first the physical and social setting:

The project I'm working on in Akko is an exciting one. We have to deal with an entire city, a walled Old City . . . a coastal city, and a city that has several layers, the most significant . . . dating from the periods of the Crusaders and the Ottomans.

Imagine! All the kings of Europe come to Jerusalem to conquer the Holy Land. For two years they rode to the Mediterranean, to the Middle East . . . with the vision that this part of the world should be Christian. And they built Akko as the port, as a starting point, like a springboard into the Holy Land. So actually you get a transmission of culture here from Europe into the Middle East.

You have here the meeting place between the universal culture, expressed by the Crusaders, bringing the Romanesque and early Gothic architecture into the Middle East, and on the other hand, you have local vernacular. This is a wonderful meeting place.

At the same time there is another dimension to this, and that's the local population: People who have been living here, or trying to make their living out of the sea or out of crafts, out of traditions.

And then there is another dimension, too, tourism, which is very interesting for me: the meeting between the local population and tourism. This is one of the major phenomena of modern life: We want tourism, but we don't want to change our way of life.

What we're trying to do is to understand the city in its many layers. It's a very complex situation, because you have to deal not only with 12th and 13th century stones, but you have to deal with people's hopes and expectations, and their poor conditions of living and unemployment.

Now, for many years, the government policy was that this should be just for tourism. It's a wonderful tourist site. But we felt that tourism cannot—and should not—be disconnected from the local population. So we've had to understand how we could actually articulate that. . . .

The concept for many years was to develop tourism just along the water line of the Old City, because tourists are interested in the water. . . . But the few thousands of people living here, were of little interest to the people . . . dealing with tourism, who said, "Let us deal with tourism, and they can do what they want! Or maybe nothing will happen and the city will deteriorate further. . . ."

We thought that there's no way to supply tourist services if you don't deal with the entire complexity of the Old City. For example, the educational system outside the Old City is much better than inside it. So we had to understand the needs of this population, in terms of kindergartens, nurseries, day care centers and services for the elderly people, and this is what actually we're working on. . . .

Now, it took a long time, a great part of my efforts, to convince the government that you can't have tourists if you don't deal with the entire population, and with the real needs of these populations. I'll give you an example.

I walked here, on one of my first visits to the harbor, and there was a fisherman drying his nets in this caravanserai, this courtyard where the caravans rested. There was a group of tourists walking around, and they walked on top of his nets. I thought that this was an expression, a brutal expression, of alienation. They may be going to the restaurant to eat his fish, but they didn't see the person, he was transparent. They just walked right on his nets, and then they went to the restaurant to get fish. And I felt he was being alienated, he was not treated as a human being, but was kept as someone who provided services, that's all. I thought that we had to respond to that.

. . . There is no high school here that meets the national standards, so we are locating the main school (and this is now approved!) right at the northern boundary—so that the population from the Old City and the population from the new city can really meet at the walls. . . . This is a social concept that responds to the structure of the Old City.

Setting out that context, Rahamimoff next described the planning and design process:

So . . . we had to convince the government to see the Old City as an entity. To do that, first of all we had to make it clear that there are no "tourist-only" services: there is no pavement for tourists which is separate from the pavements for the residents; there is no infrastructure just for the tourists, and no infrastructure just for the residents. We have to see the holistic qualities of the city.

They said in the beginning, "We only have money for the outer edge, for the waterfront, for tourism, and not enough for the inside of the city." But we said that this would be like giving headache pills to a patient who may have cancer. Because the problem is real: There is no infrastructure—the infrastructure has not been taken care of for a few hundred years. So you have to supply infrastructure for all of the Old City.

It took awhile to explain this! Now the government is in full agreement with this concept, and we're getting good money to restructure the infrastructure of the whole city—but this took not one meeting, but maybe two hundred meetings.

What happens when we meet with them? I have, I think, half a square kilometer of drawings, and they're presented, and we have a series of discussions, and I think it makes sense, because we don't have another alternative.

There is no other alternative: you can't really have hotels here near the water when there's unemployment, drugs, poor infrastructure fifty meters away from you. You won't succeed with the hotel. Now, the alternative *is* to build hotels further down the coast, but then nobody will be able to walk here at the edge of the Old City. We thought this would be disintegrating the space, the whole city and the whole region.

I think our discussions on the whole project created a different climate, a different understanding of the potentials of the Old City of Akko. With Akko on the Mediterranean, you can approach it by land, from Jerusalem and Tel Aviv and Nazareth, but you can also approach it from Turkey, from Greece, or from France, as it was historically. So there are different dimensions to the city. We're interested in all these dimensions, and . . . there's a physical expression to urban form that responds to the origins of this city, . . . which was built by the Crusaders in the 12th century, but . . . also built by the Phoenicians who came from the north before the time of Christ. It was built by early Israelites, and the Egyptians posted people here, too. We've calculated 8 or 10 major layers.

A colleague, Raphaël Fischler, and I then asked Rahamimoff about the demands of such work: "What does it take with an architectural background to keep working relationships with the government, the local population, and the municipality, to work effectively? What do you need to bring into play in addition to your architectural training?"

He replied in a way that surprised us:

Basically we have to deal with consensus-building: what we're actually trying to achieve is a broader understanding of how things should be done. We're trying to utilize our resources in such a way that we achieve a consensus so we can utilize it.

For example, the Antiquities Authority is getting money from the government for archaeological excavations here. In many cases the archaeologists are interested in finding out what was in the past—it's exciting, especially in a place like this, because every day you find all the "goodies" in the ground: they have unearthed unbelievable findings here, unearthed and also un-watered, because in the harbor here, there are about 28 archeological sites under water. There are docks, and there are sunken boats, and there is the lighthouse—very exciting things. . . .

But in most cases archaeologists are not interested in tourism. They'll tell you: "We have a plan to work here for a hundred years, so please come back in a hundred years, and we'll hand it over to you, so you can do what ever tourism you want."

Now, I'm exaggerating, because actually I'm very lucky to work with people like . . . the chief architect for preservation here, and he's very interested in Akko, and we're trying to find a way that we could utilize the archaeological findings to expose them in a controlled way to tourists, and to improve the economic basis of this population. Because if you have some sources of income, for example, for the local population that is interested in archaeology, in preservation, being tourist guides, dealing with the economic activities around tourism, it can work out very nicely.

Here we asked, "But how do you, as the consulting team leader, work between the antiquities people, the tourism people, and the local people to try to build this consensus?" He explained:

This isn't a hocus pocus situation, it's a process. I mean it's not a miracle making thing. It's a process of trying to understand the needs, trying to understand the opportunities, and trying to understand the red lines of each discipline, what's a taboo, what cannot be done, what they will not accept. There are few things that the archaeologist will not accept, there are other things that the local population will not accept, there are a few things that the tourist people and the municipality insist on as essentially important for them, and I'm interested in trying to understand each part of this complex matrix of interests.

This is something which is going on. For one thing, we have to build the confidence of the local population, that we mean work. We have to build the confidence of . . . the holding agency for the government: that economically by joining resources this will be good for everybody, that one plus one is more than two. Again, this is a synergistic concept. If you do something which is good for many of the components of the matrix, it would create a richer whole and a more meaningful entity that has a better economic base. So this is what I'm trying to do, step by step, one meeting at a time.

Rahamimoff finally summed up this kind of consensus building:

In this process, with the municipality, the tourism people, the antiquities staff, we're learning from each other. And this finally gets expressed in a scheme. But you have to share this process with other people, there's no other way. You're learning from other people, and hopefully they learn from you, and then you build a consensus, an agreement, an understanding of how things should work out. This is difficult to describe; it's a dynamic situation. It's not that you have understood what everybody wants and you bring it all into a coherent scheme, but the whole situation's changing all the time. When we presented our scheme, everybody was invited, and 70 people came to a meeting, local residents, and they had never seen a plan before.

The schemes were alienating the population, when we presented them. So there is a great amount of distrust, and you have to show your plan and hear comments and keep your eyes open and your ears open and be ready to make a change because you heard something that made sense. This is the process of building a broader consensus which is even more exciting and even more complex, and, I think, it requires changing the plan again, which I think is O.K. And if you're ready to do it, because you think it will be a better plan, then your client will also be ready to do it.

I listen, and I try to speak in more depth with some of the people that had some ideas, and I have to check it with my client, and that's how you build confidence. You show that you are listening and that you are ready to change your plans. We're ready to try and convince the authorities that there's a point there. This is all happening now so I can't show you yet how it has worked out. But this is what I have been doing for the past year.

Building Confidence and Places, Deliberative Process and Built Form

Consider what we can learn from Rahamimoff's story of his team's work in Akko. He begins by telling us about the rich history of the city—a city of Crusader and Ottoman layers, a city connecting Europe to Jerusalem on the path of conquest. He speaks of the "wonderful meeting place" of the "universal culture, expressed by the Crusaders" and the "local vernacular." Similarly, he tells us of the conflict between "tourism" and "the local population," the challenges of dealing "not only with 12th and 13th century stones, but . . . with people's hopes and expectations."

He moves quickly to detail the conflicts among today's actors: "For many years, the government policy was that this should be just for tourism. . . . 'Let us deal with tourism, and they can do what they want!' . . . Or maybe nothing will happen and the city will deteriorate further." Here he passes along to us what he has faced and heard himself, as he quotes the voices and rhetoric of those he has had to deal with as an architect-planner.

But he also tells a story of consensus building, bringing architects and archaeologists, tourism officials and local officials, together to work on school siting as well as hotel siting, and on infrastructure projects that would serve residents of the Old City and tourists too. Rahamimoff tells us that a good part of his work involved showing the government what was possible—essentially teaching his client: "It took a long time, a great part of my efforts, to convince the government that you can't have tourists if you don't deal with the entire population, and with the real needs of these populations."

He gives a vivid example of the government's lack of recognition of the local population, of the invisibility of that population, with his story of the tourists walking right over the fisherman's nets, their eyes on the old walls, eyes looking through this man and his nets, the tools of his livelihood.

Rahamimoff's efforts to bring old and new together, to gain recognition and respect for the people of the Old City, were not restricted to making moral arguments; they took actual physical form. He tells us his team had already gained approval for "locating the main school right at the northern boundary—so that," he says, "the population from the Old City and the population from the new city can really meet at the walls." Here he teaches us an architectural truth that many planners might yet come to appreciate: boundaries may not only separate entities, but join them too.

The architect-planner here speaks as well as draws, persuades and teaches as well as renders, shows as well as portrays. Along with the work of site design came the work of developing a public imagination, shaping the imagination and understanding of the public actors to see the site in a new way, as a whole: "We had to convince the government to see the Old City as an entity. . . . They said in the beginning, 'We only have money for the outer edge, for the waterfront, for tourism, and not enough for the inside of the city.' But we said that this would be like giving headache pills to a patient who may have cancer. . . . You have to supply infrastructure for all of the Old City."

This was not just a matter of presenting a good and rational argument, but a central challenge of the whole process: "It took awhile to explain this. Now the government is in full agreement with this concept, and we're getting good money to restructure the infrastructure . . . but this took not one meeting, but maybe two hundred meetings."

For designers or planners who are impatient with process, this is bad, bad news: 200 meetings may be 199 more than such design professionals want. But there is good news here too: that architect-planners like Rahamimoff are transforming not only spatial relationships but argumentative spaces and, through them, public and political relationships too: a specific government's understanding and appreciation of particular historic resources, of specific social needs of cities, of ancient stones and modern hopes all at once (Schneekloth and Shibley 1995).

This work of shaping public imagination is not just incidental, not just an artifact of responding to shortsighted government bureaucrats, narrowly interested tourism officials, or archaeologists who would prefer the planners and architects to "come back in a hundred years." Rahamimoff suggests that design professionals work not only at the abstract intersection of several disciplines, but in that deliberative, argumentative space between the actual officials who represent them. Not only do the general disciplines of architecture, archaeology, education, and economics come to bear on the conceptual problems of planning and design, but more precisely and vocally do the representatives of those fields press on the planning team. As busy, harried, friendly, preoccupied, defensive, or aggressive as they may be, these representatives of the Antiquities Authority, the Education Ministry and the local school system, the Tourism Ministry, and still others come to participate in meetings—many, many meetings. And here in these meetings, Rahamimoff suggests, he must negotiate and renegotiate any future development: its scale and character, its funds, the issues that it will honor or neglect, the integration or disintegration of "the space, the whole city, and the whole region."

Rahamimoff tries not only to create a different built form that respects past and future, the broader region and the local population, but also to create different working relationships, with "a different climate, a different understanding of the potentials of the Old City." This creation of a different climate, a different understanding—the re-visioning of possibilities—lies at the heart of the design professions, and it represents their political core. For politics involves not only the distribution of who gets what, but also, and more profoundly, the capacity to act, including crucially people's imagination of "I can . . ." or "I can't . . . ," "We can . . . ," or "We can't . . ." (Forester 1989, Marris 1982).

The design professions are deeply and inevitably political, then, because they reshape our senses of hope or resignation, our shared perceptions of our possibilities, of the ways we can (or cannot) respond to the uncertainties and the contingencies, the always precarious possibilities, of human flourishing—or what we more mundanely call "community development" (Nussbaum and Sen 1993).

This deliberative and imaginative politics of design, Rahamimoff suggests, is not pie in the sky dreaming, for the payoff is practical. He refers to the product of the hundreds of meetings, the changed understanding of the Old City and its potentials, as consensus building. This consensus building aimed, he says, at "a broader understanding of how things should be done. We're trying to utilize our resources in such a way that we achieve a consensus so we can utilize it." Here we have consensus building not for ideological purposes but "so that we can utilize it"—so that design can be realized, made real, so that the reimagining of the city can be practiced, put into built form and into a new public understanding of the city's historic resources and future possibilities too (Susskind and Cruickshank 1987, Susskind, McKearnan, and Thomas-Larmer 1999).

The lessons that Rahamimoff teaches us here involve power, the realities of interdependent actors, and the first axiom of negotiation: people who can act unilaterally may well not negotiate at all. Like many other planners and urban designers, Rahamimoff finds himself working in an entangled and interdependent world in which local and national levels interact, public and private interact, professionals and lay residents interact.

He tells us that there are no recipes here, no technical fixes; it is a slow process: "This is not a hocus-pocus situation, it's a process," and he continues, reflecting an instructive humility and persistence, "I mean it's not a miracle making thing. It's a process of trying to understand the needs, trying to understand the opportunities, and trying to understand the red lines of each discipline, what's a taboo, what cannot be done, what they will not accept."

But it is much more than trying to understand those limits, too, he tells us, for he turns to the importance of "building confidence" and showing what is possible, pointing to what dispute resolution students call "mutual gain" (Lax and Sebenius 1987): "We have to build the confidence of . . . the holding agency for the government: that economically by joining resources this will be good for everybody, that one plus one is more than two. Again, this is a synergistic concept. . . . So this is what I'm trying to do, step by step, one meeting at a time" (Raiffa 1985, Susskind and Cruickshank 1987, Kolb 1994).

This showing, this teaching, is interactive, not isolated, and certainly not a one-way street either. Rahamimoff has described a negotiated learning process in which he is simultaneously an interested party, a facilitator, and a process organizer as well—and so he powerfully refutes both common stereotypes and academic presumptions that these distinctive roles cannot feasibly be combined. He says: "You're learning from other people, and hopefully they learn from you, and then you build a consensus, an agreement, an understanding of how things should work out. . . . It's a dynamic situation. It's not that you have understood what everybody wants and you bring it all into a coherent scheme, but the whole situation's changing all the time."

The architect-planner does not pretend to boil down the creative process to a recipe, but he keeps before us the promise of public deliberations (Schneekloth and Shibley 1995). He speaks of the ways he changes in the process too. "You learn," he says, speaking of himself, and they too learn, and with new information and understandings, he tells us, the parties can generate new design solutions that no one imagined before coming together (Kolb 1994, Hoch 1994). Notice that this happens *not* because of the superior understanding or analysis of the design professional alone, but because of the deliberative learning process itself: "It's not that you have understood what everybody wants . . . the whole situation's changing all the time."

The situation changes, the planner and designer listen, and, Rahamimoff tells us, the plans must change too. He seems to stress the ability of the designer to learn, to be able to reframe options, "because you heard something that made sense" with "your eyes open and ears open" (Schön 1983). Again this prospect of learning suggests a change in both the consensus that can be achieved and, more important, the prospects for implementation. This architect-planner is rebuilding a city and "building confidence" too: "I listen, and I try to speak in more depth with some of the people that had some ideas, and I have to check it with my client, and that's how you build confidence. You show that you are listening and that you are ready to change your plans."

Listening in deliberation is crucial but insufficient, for if listening does not lead to subsequent action, to the possibility that what is heard can actually make a difference, then such listening becomes merely condescension, wasting or manipulating others' time, an act less of taking the other seriously than of insulting them by failing to respond to their deeply felt concerns.

Rahamimoff suggests the moral challenges of deliberation here. Listening to the criticisms and ideas of affected people implies the potential recognition of new concerns, the re-cognition and re-thinking of value, strategies, consequences, and implications (Barber 1984; Gutmann and Taylor 1992). But all this is done "contingently," without guarantees, as matters of effort and direction: "We're ready to try and convince the authorities that there's a point there. But this is all happening now so I cannot show you that this is how it has worked out. But this is what I have been doing for the past year."

These problems are hardly unique to Akko, to the Middle East, or to the challenges of reconciling archaeology, tourism, and economic development. Like many other planners and urban designers, Rahamimoff has faced multidisciplinary problems and multiple actors with multiple interests, time-tables, budgets, and access to power. To get something done in a world where his client cannot act unilaterally, blind to local regulations and the cooperation required of others, Rahamimoff, like other planners and designers, needs not only to design forms and spaces, but also, and as important, to build confidence, shape the attention and imagination of others, and design a practical, deliberative process of collaborative—essentially mediated or facilitated—planning and design. Just as Rahamimoff has spoken of a consensus-building design process animated by hundreds of meetings, hundreds of deliberative, argumentative occasions, we need to understand still better what roles planners, architects, and design professionals can play to make such meetings work: to foster public imagination, public recognition of parties and issues, learning by all involved (Hoch 1994, Bryson and Crosby 1996).

The Sketch as a Way of Asking Questions

We can begin to examine such work more closely by considering part of a city planning director's account of a multiparty process planning for the complex waterfront of Oslo, Norway (Forester 1994a). The case involves environmental issues, shoreline land uses, and recreational boating—or, more generally, conflicting uses, scarce resources of land, water, and environmental amenity, and conflicting interests of parties with their own ideas of how best to use those limited resources. Although the land use questions here involve shoreline issues that the landlocked among us rarely experience, the challenges of mediated negotiations, deliberation, and the realistic design of workable options are all too familiar.

This director, Rolf Jensen, used his staff quite deliberately as mediators to manage special planning subcommittees. He tells us:

> Sometimes, we would set up smaller groups to solve special issues: we'd put them alongside the planning group as "specialist groups," as we called them. It's a good name—although there are common, non-professional people in them sometimes—it gives them also the feeling that they're doing something good. . . .
>
> For instance, there's a small island here in the middle of the harbor on which they store boats. There're some services for boats; there're sailing and recreational boats here, but there's also a concern from the house owners, along the shoreline, to limit that, so it won't sort of creep into their land use. And there're also some environmentalists at the county level who were concerned that there was too much boating down in this area: it disturbs the birds on some of the smaller islands out here—and things like that.

This was an issue that was hard to solve. So we created a special group, trying to come up with schemes for this area, and then the planner would be just a mediator in that group. The planner would let the parties argue, and try to find solutions; they would work with colored pens and papers; they could write; they could do whatever they liked. They had what you might call workshops together, in which the basic task of the planner was to get the parties to understand each other—because in the Norwegian tradition, many times, you just present the maps, and that's it: "Take my demand or not!"—a sort of power play.

We tried to conceive from the first day that we are here to listen. We are here to try to understand. But we are also here to try to tell you a story—in other words why we are concerned about certain things. That's classic in negotiation. You try to tell them more about your interests—why are we putting forward these demands? Because if you do that, you gain two things.

First of all, the other party recognizes you too as a party—because you have, naturally, some reasons for what you want. It's conceivable to them that you would have that set of reasons behind your demand. That's the first one.

But also, secondly, you might be able to help that party to come up with other demands.

This happened both when we as planners met with individual groups and met altogether—all the time! That attitude we used over and over and over again: the attitude of never presenting a sketch as *the* sketch. Always saying that, "Look, the sketch is not important, but what I've been trying to find a solution to, through this sketch, is this and that and that and that and that and that." In other words, it was the intentions and the characteristics with the sketch that was important, not the sketch itself.

It was important as a way of asking questions, and as a way of controlling questions to the parties: "Does that serve your needs?" "Is this something that you can live with?" Or, "What is really burning you if you look at this sketch?"

Sometimes it could be small things. In one instance, we had boat piers sticking out from the shoreline. We found in this one area that as long as those were not longer than, let's say, thirty meters, it was all right. If they were longer, say forty or fifty meters, then immediately we would have groups objecting to it. Because the way that they understood their ability to maneuver their boats and get around in the area would be basically much more limited if the piers came out that far.

So, then, it wasn't the question of "piers or not?" but the length of the piers. See?

As Questions Frame Responses, So Analysis Shapes Design

In form and function, this planning subcommittee resembles many others with which planners work quite routinely. Citizens' group and agency representatives meet to work out their differences and propose workable options—for a city council vote, a mayor's decision, a community organization's membership, or an official board's vote. The planning staff bring technical information and, as Jensen makes clear, sometimes a good deal more too. Meetings like this, as Rahamimoff suggests too, are quite ordinary; they happen all the time. But this quite ordinary example of a local planning subcommittee is extraordinary too, for it provides a microcosm of successful planning efforts to encourage public deliberation.

Jensen is not just telling us about a trivial case, for he begins by saying, "This was an issue that was hard to solve." Stuck on this land use issue, Jensen created the subcommittee, as he says, "trying to come up with schemes for this area," and he designated a particular role for his staff planner: not only as a source of information and expertise, but as a mediator, and not simply and distantly as a neutral mediator, but as a mediator with a mission—as he put it, to "let the parties argue, and try to find solutions."

Jensen's allusion to "colored pens and papers," to having "workshops together," in which "they could do whatever they liked," suggests a range of possibilities more than any one strict structure or procedure, but he also suggests a "basic task": "to get the parties to understand each other," "understanding" here not to be "nice" but to get practical work done.

All this might sound ordinary enough, but Jensen explains that this mediating role, this deliberate fostering of argument, this search for solutions together was hardly the norm. "In the Norwegian tradition," he explains, echoing our own sense of power-politics in the United States, "many times a party might just present the maps, and that's it: 'Take my demand or not!' " take it or leave it, "a sort of power play." Hardly blind to questions of power, Jensen's story helps us to reimagine planners' roles when power is ever present—but not so unilaterally strong that there is nothing to negotiate at all (Flyvbjerg 1998, Forester, forthcoming).

Jensen suggests that the mediating planner is not just a neutral facilitator. This mediator-planner instead seeks solutions ultimately, and a host of lesser objectives along the way: "We are here to listen . . . to try to understand, . . . but . . . also to try to tell you a story . . . why we are concerned about certain things." Like Rahamimoff, this mediator too is a full-fledged participant—not only a manager of a process but a negotiator and a mediator at once: a negotiating partner seeking to facilitate a process as well as to explain substantive concerns or "interests."

Jensen's mediating planner was there, he says, "to tell them a story," not to be cute, not to be entertaining, but to do work (Innes 1996, Mandelbaum 1991, Throgmorton 1996), and very interesting work at that: to move not only from positions or particular solutions to "interests" or underlying concerns, but to do still more too.

Jensen explains that the importance of planners' telling their stories about their concerns involves the recognition of the other parties: "The other party recognizes you too as a party—because you have, naturally, some reasons for what you want."

The planners' storytelling does a great deal more than describe events, establish debating points, or even clarify underlying interests. That storytelling helps to establish a relationship between the parties, and not a relationship of intransigent adversaries but one of reasonable differences: "It's conceivable to them," he goes on (and we might say, "imaginable" or "reasonable" then), "that you would have that set of reasons behind your demand" (Benhabib 1992, Calhoun 1994, Gutmann and Taylor 1992).

Here we see that as planners tell stories, those stories simultaneously tell a good deal about the planners too. We all tell stories, and our actions create new stories; both of those stories can then disclose and reveal our concerns and our character too (Arendt 1958, White 1985).

Jensen is telling us that even more can happen here too, for the planners' stories might "help the other party to come up with other demands." This might sound counterintuitive, but it becomes more plausible, as Mansbridge suggested, when we remember that the parties in planning processes often have limited information and expertise, still less trust and understanding of one another, and, unfortunately, keener views of what they fear than of what they might really achieve together (Taylor 1998, Innes 1996). Under such politically cloudy conditions, when the potential relationships and agreements between the parties are even more ambiguous and uncertain than the "facts of the case," the planners' carefully told stories of their concerns can suggest options beyond those initially imagined by any of the parties at the table.

Jensen's portrayal of the mediating design professional involves still more than storytelling, relationship building, and information providing. He asks us to think of the planners' and designers' sketches less as representations than as questions, as probes, as visual ritual objects designed to elicit issues, ideas, suggestions, and proposals for further refined options (cf. chapter 5 below, Healey and Hillier 1996). He tell us in a wonderful passage, "That attitude we used over and over and over again: . . . never presenting a sketch as *the* sketch. Always saying that, 'Look, the sketch is not important, but what I've been trying to find a solution to, through this sketch, is this and that and that and that and that and that.' In other words, it was the intentions and the characteristics with the sketch that was important, not the sketch itself." And then crucially he continues to say that presenting a sketch "was important as a way of asking questions, and as a way of controlling questions."

Jensen teaches us a good deal here about the challenges of design professionals in playing mediating roles, roles fostering deliberative processes in which parties can learn together about one another and about their joint possibilities. Jensen's planner-mediator is a designer who needs to learn and knows it, a design professional using his or her ability to sketch as a way of investigating and learning about both fact and value at once. This planner does much more than a conventional "process manager." Jensen's mediating planner uses professional skills to explore both "values" (the "intentions" of the sketch) and "value" (what might matter, what might emerge as important in the case), the "characteristics" Jensen mentioned being rendered by the sketch. Stressing the importance of learning together, Jensen emphasizes the planner's role of asking questions (Forester 1994b, Susskind and Cruickshank 1987). This managed learning process is not an automatic, natural, or mysteriously creative one, though, and Jensen does not minimize the importance of control, the sketch as a way of controlling questions, as well.

Even Jensen's questions suggest a good deal about what the planners and subcommittee members need to learn: "Does that serve your needs? Is this something that you can live with? Or, "What is really burning you if you look at this sketch?" These three questions map important differences, all of which must be explored in public deliberations. If the first question explores a party's more ideal conception of the needs it wishes to fulfill, the second question takes a more strategic and practical turn: Can you live with this option as presented? Is this sufficient, even if not ideal?

The third question too is crucial to deliberative processes: "What is really burning you if you look at this sketch?" This question explicitly addresses emotion, and it concerns emotion for a reason. Jensen is telling us, as Martha Nussbaum (1990a) has, that the emotions are potentially modes of vision onto the world, that if we can learn about what is "burning" someone, we are likely to learn more about what really matters to them, more about what else they may care about, as well as what they are likely to fight to protect or resist. We can learn not only about the emotion, then, but about the practical and often malleable world that has engaged it. Learning through emotion, we can learn about aspects of solutions and options that might yet be changed, fine-tuned, altered, and redesigned.

Such learning through emotional response is barely appreciated in our public decision-making texts, even if that is less true of a growing negotiation literature. But here practice leads theory, in planning and public decision making, by light-years. Jensen is asking us to train planners not just to know about abstract solutions, but to pay attention sensitively to citizens' fears, anger, suspicions, and their related emotional reactions to the planners' only apparently innocent and well-meaning sketches. Far too little planning literature addresses these issues (Nussbaum 1990a, Susskind and Field 1996).

Jensen is interested not in theories of rationality here, but in practical results, as he goes on to illustrate. Citing the controversy over the boat piers, he tells us how the process successfully reformulated the issue at hand: "We found in this one area that as long as [the piers] were not longer than, let's say, thirty meters, it was all right. If they were longer, say forty or fifty meters, then immediately we would have groups objecting to it."

His point, significantly, was that the subcommittee had learned about the real issue before it. They had reformulated their sense of the problem and their sense of the solution too. Their deliberative practice, reconsidering land uses and amenities and environmental concerns, led them to a new appreciation of how they could go forward together (Hoch 1994, Susskind and Cruickshank 1987). As Jensen summed it up, "So, then, it wasn't the question of 'piers or not?' but the length of the piers. See?"

We might make less of this Oslo subcommittee if planners and designers did not spend so much of their time in meetings like this. But gone are the times when planners could work in isolation and expect a pleased public decision-making body to accept their work with gratitude and implement it straightforwardly. Today, instead, we have planners particularly working in between interdependent and conflicting parties, in situations calling for a good deal more mediation, facilitation, and collaborative problem solving than for unilateral, top-down or one-way, planning action (Susskind and Field 1996, Ventriss 1991).

Jensen's Norwegian planners face challenges quite similar to those confronting Rahamimoff's team in Akko—challenges we face in the United States all the time too. Jensen's simple story of the planning subcommittee illuminates the complex challenges of the participatory processes and public deliberations that design professionals can encourage. Those planners and designers will need not just to manage process rules but to foster argument. They will need to bring their own tools to bear to explore issues and raise questions, to learn not only about strategies and means, but about what matters to people—what is at stake, too. They will need to build confidence and relationships as well as to present credible information and expertise. They will need to explore emotions and arguments, always seeking to facilitate a consensus about practical solutions.

Educational and Theoretical Implications

We can close by giving the last word to a Washington, D.C.–based planner and mediator, William Potapchuk. Trained in urban studies and planning, Potapchuk facilitated an ambitious yearlong process to rewrite the zoning ordinance in quickly growing Loudoun County, Virginia. Toward the end of his account of that work, Potapchuk pointed to the skills and knowledge necessary to those entering the public dispute resolution field (Forester 1995). He spoke first of a core knowledge of the political system in which planning is embedded, but he pointed to six further requirements for planners as well:

> In having been part of several hiring processes for a number of positions in the [dispute resolution] field over the years, what we've discovered in general is that while you can teach the array of dispute resolution and consensus building skills

that are necessary, especially for people who've had some background in the field, you cannot teach political understanding and political sense. And the core knowledge is understanding politics and how it works: the interplay between the governance process and the electoral process, the role of staff and advisory bodies and how they work, and how the lobbying process works. Unless you understand the system as it works, it's going to be very difficult to act as an intervenor and as an advocate for changing the system and changing the process.

So once you start with that core understanding and know what you want to change, or what the difficulties are, and what the opportunities are, then, first, you have to have that skill as a negotiator. Ultimately, you're an advocate for a different kind of process, and that advocacy plays out a number of different ways, and you need to role-model good negotiation behavior.

Then you need facilitation skills, your ability to manage a large group of people using a variety of collaboration and problem-solving tools to help a group identify issues, identify options, and come to agreements—not to be a one-trick pony, but to know a number of different approaches to making that happen.

Thirdly, then, you need meeting design skills. What kind of meeting do we need for this kind of task?

Then, fourthly, process design skills: How do we structure a series of meetings over an extended period of time to accomplish broader tasks, and how do we know whether this is going to take two meetings or six meetings, and how many issues might emerge? How do you begin to figure that out so you can craft a yearlong process?

Then, underlying all those tools, you have to have a good, clear sense of mediation strategy. I've often thought that what we do as intervenors looks more like facilitation, but uses the strategies of a mediator. You have to understand how to bring people to agreement, how to nest the process in the larger political system, how to maintain the connections with key actors, and so forth.

Then lastly, you need some substantive understanding of what's being talked about. People talk about different elements of a zoning ordinance. What are those elements? So when a member of the task force or the community or the staff throws out an acronym or throws out a section, you know what that is, and you can react accordingly without either missing something, which is very dangerous, or continually having to slow a group down and say, "Well, what are you talking about?"

I do think in terms of planners and planning. . . . One reason that I'm a member of the American Planning Association, and I stay active professionally, is that the profession largely does not recognize the role of planners in making public processes work . . . yet if you go to work for any kind of community where you're on the front lines, as much as fifty to seventy percent of your time is spent on designing processes, participating in processes and making meetings work.

Potapchuk concludes then: "Now obviously that doesn't mean that you enter the field not understanding basic planning principles, but that planners are often the guardians of democratic processes on the local level. They are often the key staff people who are called upon for making public processes work, and yet they receive very little training in the area. I think that argues for a change in the curricula that planners receive—and for a change in the profession."

Consider five further points. First, Rahamimoff and Jensen raise issues of process design (and more profoundly political and institutional design) that are barely discussed in the planning literature and only briefly treated in broader analyses of consensus building and dispute resolution (cf. Ury, Brett, and Goldberg 1988, Dryzek 1995). Such issues arise whenever planners must organize a series of meetings, devise a process to review a major proposal, or recommend a new procedure to be voted up, down, or sideways by a planning board or a city council. Such questions of political design are those of creating the political and institutional spaces in which public argument can flourish, educating and improving relationships between interdependent parties instead of simply escalating hostilities and deepening the resentments of such parties. These challenges of political design are practically commonplace but all too theoretically ignored in the planning literature (for exceptions, see Innes 1995, Healey 1997). Here, then, the insights of skillful practitioners set an agenda for the future theoretical work we must do, if we are to do justice to the real demands and aspirations of a critical planning practice.

Second, this analysis reveals the politically deliberative character of consensus building and mediated negotiations as they arise in many phases of planning processes. Political boundaries join as well as separate; public officials, citizens, and experts can argue, listen, and learn over the course of many meetings. Without an understanding of both the potentials and difficulties, though, those promoting deliberative design and planning efforts will risk celebrating narrowly pragmatic deal making instead (Schneekloth and Shibley 1995).

Third, this analysis can help us to see that just as conflict among parties with differing interests in planning is commonplace, so too will the overlapping skills of negotiation, mediation, facilitation, and consensus building be commonly in demand as well, especially in the absence of third-party intervention by special "process consultants." But how might these skills complement one another? Academics announce conflicts between these roles, but design professionals must often combine and integrate them creatively nevertheless, for example, by negotiating and mediating at the same time (Forester 1989). Good planning theory should clarify how that is possible, for better and worse, and not just tell us that such roles conflict.

Fourth, by complementing the recent work of Charles Hoch (1994) in Chicago and Patsy Healey (1997) in the United Kingdom, this analysis helps us to see that carefully crafted deliberative discussions are realistically possible in adversarial contexts. ("Where else?" the politically astute would ask!) Thus, the temptation to distinguish planning settings and situations as adversarial *or* collaborative is simplistic, misleading, and self-defeating. The challenge of democratic deliberation is not to avoid, transcend, or displace conflict but to deal with practical difference in and through conflictual settings (Sandercock 1995). Complementing our planning literature (e.g., Susskind and Ozawa 1983, 1984, Susskind and Cruickshank 1987), Rahamimoff, Jensen, and Potapchuk's remarks help to show how planners can promote such deliberations in concrete cases.

Fifth, we see here the intertwining of process and product, neither new technical fixes nor recipes, but a range of practical considerations that come into play in real cases. At stake is not just consensus building, but the integration of acting and learning, relationship building and world shaping, that reaches far beyond narrow deal making to the creative practice of deliberative planning and design in the public sphere.

4

Recognition and Opportunities for Deliberation in the Face of Conflict

For several years as I have worked on this book, I have heard the echo of an Israeli transportation planner's words. I had been asking, as I usually do when I can interview planners, about how he worked on difficult public projects in the face of great uncertainty. How did he work when many parties were invested, in conflict with one another, and not often pleased with their prospects? I was interested too in the issues reflected in the title of this chapter: How could planners in messy, political situations foster public deliberation, not just deal making? How could they encourage public learning, not just posturing and inflated rhetoric? How could they improve public decision making, not just strengthen the hand of the already powerful? For his part, looking back at twenty-five years of experience since his MIT training, this planner was speaking about the challenges of developing productive working relationships in the face of conflict. He was telling me that November afternoon about the importance in his everyday work of what he called "diplomatic recognition."

At first I thought that this planner's story and his insights would be peculiarly limited to the Israeli context. But I was largely wrong; the broader relevance of this planner's professional experience in contentious meetings has become much more obvious to me since I returned from Israel.[1]

This chapter explores the challenges posed in fostering democratic deliberations, the work that I take to be the promise of planning. By "democratic deliberation," I refer to the practical public imagination of the future in a variety of real decision-shaping discussions, in community meetings or negotiations, involving either representatives of public constituencies or directly affected citizens themselves. By "deliberation," following Aristotelians like Martha Nussbaum and critical theorists like Seyla Benhabib alike, I refer to conversations that involve more than the evaluation of efficiency—assessing which options, strategies, or means provide the most bang, the most social benefit, for each buck (Nussbaum 1990a, Benhabib 1992, Mansbridge 1992, Dryzek 1990, Lindblom 1990). Political deliberation also involves, as we shall see, two more complex and challenging kinds of practical work: a careful exploration to learn about ends (including goals, mandates, obligations, hopes, and what these mean in a given case) and a subtle but real recognition of other parties—even as *they* might propose to build where *you* want to preserve (or vice versa), even as they bring histories of distrust and feelings of being "done-to" to the table.[2]

To explore these issues, we can consider recent research and planning experience in the United States, Europe, and the Middle East. As different as these contexts are, we can learn from all of them, in particular by exploring three accounts of deliberative practice given by an urban city planning director in Olso, Norway, an architect-planner in Tel Aviv, Israel, and the transportation-consultant-planner mentioned above, working in Israel.[3] These practitioners can teach us a good deal about the daily professional and political challenges of encouraging public deliberation, and improving public decision making, in and through astute planning and design practice.

Meetings, Meetings, Meetings—Not Whether, But How to Deliberate in Adversarial Planning Settings

Before assessing material from Europe and the Middle East, consider first recent research from Chicago. In a recent book exploring the work of planners in and around the city, Charles Hoch argues that we should understand democratic deliberations not narrowly and nostalgically as anachronistic discussions in New England town meeting halls, but more broadly to include the always precarious practical, public discussions that span institutional boundaries and involve the multiple parties and shifting coalitions with whom planners typically work (Hoch 1994). If his analysis is even half-right, the practical question for planners and design professionals in a political world is not whether to be either adversarial or deliberative, but *how best* to be deliberative within conflictual, adversarial settings (Lax and Sebenius 1986; chapter 6 below). Planners are involved in complex public deliberations all the time, Hoch argues, but we know too little about how they can improve them.[4]

Relatively few people take seriously the model of the planner as a politically authoritative expert whose legitimate power is ensured solely by his or her special substantive knowledge. Specialized knowledge is only one source of authority, whether it concerns zoning codes or housing strategies, urban design methods or historic preservation strategies, economic development or environmental regulations. The truism about the planning and policy process, "It's all political!" increasingly becomes a self-serving excuse for inaction rather than an explanation. We know that planning is significantly political, that planners' recommendations are interpreted and implemented for better or worse in political processes, that planners may at times study only certain options at the request or pleasure of politicians, and that the distance between rational public policy and political will can be substantial. No one has disputed any of this in the planning literature for twenty, if not thirty, years. We know too that the perpetual rediscovery of politics can easily turn out to be more complaint than insight, especially when it tells us little strategically and practically about what planners can do—how they can still be able to improve public decision making and the resulting public welfare. To tell us that planning is political without presenting a clear analysis of what that implies for effective practice is simply to substitute the label *political* for the hard work of figuring out what can best be done. As Paul Goodman put it wonderfully, the question is not optimism or pessimism, whether planning is political or not, but *how* to act, or, as he wrote, "What now?" (Goodman 1951:xiv, Forester 1999a, forthcoming).

A few years ago, a simple peace was made with the politics of planning by thinking of planning as advising decision makers. But our sense today of the politics of planning is far more complex. We know that the "advice" in "advising decision makers" is never neutral, that decisions are taken from prestructured agendas as much as they are "made," that citizen access and political responses are influenced by wealth and organization, by class, race, and gender, that the very language in which we pose and discuss problems can be politically selective, inclusive or exclusive, and that planning can be influential in predecision processes involving citizens' education, mobilization, and subsequent negotiations (Friedmann 1987, Forester 1989, Grant 1994, Marris 1982, Yiftachel 1995). The recent explosion of interest in dispute resolution and mediated negotiation in planning, for example, points to an area in which planning influence is no less political, but is nevertheless directed not toward traditional decision makers as much as toward the staging or even organizational design of public negotiating processes (whether these involve land uses, site planning, or environmental regulations, for example) (Susskind and Cruickshank 1987, Susskind and Ozawa 1983, 1984, chapter 6 below).

Hoch's book takes us far beyond the old image of planners processing information to decision makers who subsequently make professionally informed decisions (Hoch 1994). Hock argues that a politically informed pragmatism can give us a more realistic, both rational and political, understanding of planning practice: "The pragmatists replace the model of the planner as an expert offering truthful advice to the public with that of the planner as a counselor, who fosters public deliberation about the meaning and consequences of relevant plans with those who will bear the burdens and enjoy the benefits of purposeful change" (1994: 294).

A "pragmatic approach" to planning, Hoch says, "emphasizes the politics of deliberation" (1994:298). But we should ask, as political skeptics and as realists who have inevitably been burned as idealists, "Is this 'deliberation' simply the romantic exception that proves the rule of nasty, brutish, and short adversarial politics?" Will the "deliberation" that Hoch finds so promising mean simply that while planners encourage more inclusive public voice, the powerful will continue to decide as they wish?

Hoch asks us not to be so simple or cynical in our skepticism, not to be so dismissive that we miss the opportunities that planners do have. The very messiness and indeterminacy of urban politics produce such opportunities in planning, he argues: "Political opponents involved in complex and longstanding conflicts surrounding such planning policies as land use regulation, energy conservation, public housing, and public health cannot withdraw like prize fighters to their corners to get relief, but they *must* carry on in public, seeking compromises and mutual support. When adversarial relations prove exhausting, perverse, or inconclusive, then the prospects for deliberation improve" (Hoch 1994:302, cf. Krumholz and Forester 1990, Forester 1998b).

Hoch really makes two important points about deliberative opportunities in planning. First, as I have suggested, he asks us to rethink the relationships between the adversarial and deliberative aspects of practice. He shows that we get it wrong by forcing an either-or choice, even as we know that adversarial strategies can preempt and prevent more productive deliberative discussion.[5] Hoch asks us to develop the double vision that students of negotiation consider second nature: the ability to pay attention not only as narrower "technicians" to the "substance at hand" (what we are negotiating about), but also to the "relationships between the parties" (how *we* are negotiating this, how our mutual suspicion of one another may be tying our hands so we both do poorly, how the implicit design of the process—the ways we meet—rewards exaggeration and escalation rather than more creative joint problem solving, planning that might satisfy real needs). Just as negotiation analysts explore what they call integrative or mutual gain possibilities within apparently distributive or zero-sum disputes (Susskind and Cruickshank 1987), and just as democratic theorists explore what they call transformative possibilities within what seem at first to be only occasions for simple bargaining (Barber 1984, Kemmis 1990, Mathews 1994, Warren 1992), Hoch challenges us to learn these lessons in planning so we can engage in adversarial and deliberative action together.[6]

Second, Hoch makes a more subtle point about planners' opportunities to shape public deliberations among interdependent parties. When one party to a planning question or dispute—a neighborhood association, a developer, or a state agency—is powerful enough to act unilaterally to get what it wants, there may not be much to talk about, nothing to negotiate. Planners will have no influence; that is what the other's ability to act "unilaterally" means, after all. But we should not let these often-high-profile cases obscure the great many others that planners face all the time—cases where planners have important roles to play by bringing together uncertain and hardly omnipotent parties—project proponents and opponents, regulators and neighbors, local and state agencies—who are interdependent, interconnected, affecting each other by action or inaction.

Hoch tells us that when the parties to planning disputes are interdependent, when because of uncertainty, limited resources and power, and shifting relationships the parties need to come to terms practically with each other, then planners can have practical opportunities to shape that "coming to terms," those discussions and negotiations, as they take place.[7]

Here planners' and designers' practical influence extends far beyond providing information to decision makers. Planners can make a difference not just by controlling information and setting agendas, not just by shaping public arguments and framing options selectively (Gaventa 1980, Schön 1983, Throgmorton 1996), but in more subtle ways too. Planners can also exert real influence, Hoch suggests, by shaping processes of inclusion and exclusion, of participation and negotiation—in the processes of environmental or design review, for example. Planners, he implies, are already practically involved in (and they need to know much more about) questions of organizational and political design, for example, that most planning and design schools barely address. How does the deliberate structuring of review, negotiation, and planning processes shape collaboration or escalation, mutual learning or polarization? How does the design of planning processes shape public attention and neglect, filter some issues in and others out, respect and recognize some differences and not others (Lukes 1974, Dryzek 1995, Elkin 1987, Michelman 1988, Sandercock and Forsyth 1992)?[8]

Hoch challenges planners and design professionals to rethink the significance of "all those boring meetings" they have to go to—both as they participate practically in those meetings and as they stage and structure, and design them too. Unfortunately the planning and public policy literature has not assessed these problems very well. Too much of the negotiation and mediation literature, for example, remains economistic, more concerned with trading and exchange than with learning, more concerned with interest-based bargaining and "getting to yes" than with the broader public welfare and the practical and political significance of public deliberation (chapter 6 below, Menkel-Meadows 1995).

Hoch put it this way: "Analysts [of planning] have tended to overemphasize the scope of adversarial relationships in planners' work and to underestimate the social significance of their technical activity. The analysts have fostered commitments to distinctions that not only miss the complexity of planners' activity, but also offer categories of moral interpretation that blind practitioners to the importance and value of their actions" (1994:315).

He goes on to echo and develop further arguments about planners who played multiple roles—being negotiators for municipalities or nonprofits *and* mediators, process managers, and public managers alike. Hoch summarizes the wide experience of the Chicago area planners, whose stories he presents, this way: "Professional planners work in an institutional order of competitive and hierarchical relationships, which, despite their adversarial and instrumental qualities, require some cooperation. Many professional planners regularly try, in imaginative, incremental, and occasionally grand ways, to shift attention from the adversarial to the deliberative. Their persistence testifies to the effort they are making in our competitive liberal society to keep alive the practical possibilities and the hope of responsible, free, and informed deliberation" (1994:337).

Here we find the promise of planning, Hoch suggests, in bridging expertise and participatory processes, integrating rationality and politics, remaining as sensitive to issues of power and access as to questions of design and economic analysis. "Planners," he tells us as he tries to do justice to his title, *What Planners Do,* "are not only gate-keepers, but path-breakers; not simply visionaries, but counselors; not just power brokers, but public servants; not only experts, but teachers." So, he argues, "The effectiveness and practical meaning of what planners do might be better understood if judged as efforts to establish, expand, and refine . . . public democratic deliberations" (1994:315–316).

Hoch asks us to refocus our vision and look at planners in a relatively new way—as designers, managers, and even leaders of public deliberations. If we are to do that, we need a better understanding of the demands of any public deliberation that will mean more than "the old boys talking in the smoke-filled back room." To begin to explore these issues, we turn to three thoughtful practitioners' accounts of planning and design practice, accounts that can teach us about the challenges of managing interdependence and improving public deliberations at the same time. Let us ask, as we go, what this work requires of planners and designers—and what it promises as well.[9]

The Precarious Deliberative Work of Planners and Designers: Three Accounts

A City Planner's Harbor Negotiations: Participation and Deliberation in Oslo

Consider first one of the city planning directors of Oslo, Norway, Rolf Jensen, quoted briefly in chapter 3 and who, shortly after this interview, spent a year in Jerusalem heading a special Norwegian-sponsored land use planning team for the West Bank. After speaking about his background as a practitioner and teacher, Jensen describes a negotiation and deliberation involving the massive Oslo harbor. He began:

> What I learned from my early work with citizen participation was that if you could get the public truly involved, not only interested in answering or commenting on ready-made plans, but if you could get them really involved, you would gain two things. First of all, you could gain an interest in what was happening in their community, which was not present all the time. Sometimes the Norwegian people are brought up thinking that public services are some sort of a natural, God-given thing. You just sit back and wait for it, right? But as things evolve in society, that will not be the case. For instance, regarding a small local sports arena, you'd have probably to build it yourself. And indeed, in many places that happens. That's the first thing—people would feel that it was in their own interest, and they would feel proud of doing things like that.
>
> Second, the solution itself might in fact be better because they knew exactly what they were going to use it for—several purposes, for instance, combining things like skiing in the winter and football in the summer, or soccer as you call it. Whereas if you just design the area for soccer, which was probably what the planner would do, without thinking about it, then it might not be as useful in winter time, depending where you placed it in the landscape, for instance. So when you work with people you could immediately get a better solution, many times, than just doing it the "professional" way. We saw a couple of examples of that, in fact, in practice, in small municipalities.

Jensen's interest in participation developed at the expense of fighting the planning establishment:

> But anyway, all through the time I taught at the University I also developed an interest in more negotiative processes through actually working with planning

theories. I was opposed to the planning profession in Norway in another respect. Through my background in Berkeley and the United States I really got to believe in a sort of pluralistic society. I saw advantages for that sort of society. I saw also advantages in a more or less stepwise progression in planning—not fully Lindblom's incrementalism, but I liked the arguments of Lindblom, even the philosophical ones of Braybrooke and Lindblom's book: "Who am I as a professional to know what's right for you?"

There were practical arguments there all the time, but they tied in to theory. Theory was not popular with the planning profession in Norway at that time, because there was *one* way of doing planning, and that was according to the law. The law was really a super-rationalistic type of law, saying that you should start with the overall and then go to the small and that the people could come in and say what they would like to comment upon, and you could listen and if it was a good comment, you would take care of it, and if it was not a good comment you wouldn't, right? It was a top-down approach, all the time. That was *the* way of doing planning, and if you didn't do that planning, you did bad planning, poor planning. Now, this is a bit over-simplified, but nevertheless true.

Jensen suggested that such top-down planning was not working well:

I reacted against that all the way. And I could prove my case as time went by, because when we hit the '70s and the first general [municipal land use] plans then came out (following the brand new planning law of '65), you could *see* that they were dead plans, they were hardly ever used. In fact the only reason many municipalities made their general plans was in order to get state funding. And you could show that, over and over again.

And of course what did that lead to? Many of the planners quit the profession in the mid-70s. They couldn't stand it. They said that if that is going to be planning, it's not the place for us to work. Because many of them had an almost religious belief in planning. They wanted to *help* the land develop, they wanted to *help* the people and so on. But then they saw the results, and they felt that they had fooled themselves or that they had been fooled by the system.

This all fed my interest in negotiation. The incrementalist planning theory also leads toward negotiation. How else can you then do the step-wise work, and how could you otherwise cope with pluralistic wishes? You cannot cater to all of those, you *don't even know* what they are; you have to look for them in some way. And how do you look for them? By getting into dialogue with people—and as you get into dialogue, you also enter into negotiations.

Furthermore using questionnaires and then planning did not lead to the good results that were expected. Many of the planners also realized that, in fact, they played the role of politicians sometimes: that many of the things we took for granted in planning—for instance some of the standards, say for highway transportation the curve radii that were to be designed—were actually normative decisions, political decisions, because they said something about standards and how much money you had to put in to achieve them, and the money then would be taken from schools and other things, right?

So they said, "Boy, we ought to be careful here; we ought to be careful. This isn't a kind of natural process at all. We need more judgment from outside to be sure that we are really doing a good job."

And how do we get that judgment from outside? Again, by dialogue, and I would say not by downgrading your own profession, but by downgrading the image of your own profession, having more professional humility—reflecting first of all upon what you're doing and being more open for more negotiated approaches."

That sounds admirable, but what does this mean in practice? Jensen elaborated by describing two cases he had worked on, both concerning harbor and waterfront planning and negotiations (we consider only one here) (Forester 1994a). He explained:

So when I came to the Oslo City Planning Office I came first of all as a professor, which in itself was dangerous. They would not view me as a practitioner, right? And since I had been the only one in Norway giving courses in planning theory and advanced planning methods, they might have viewed me as a threat: "Here comes someone from the outside, telling us what to do." So I had to build confidence. You can do nothing without confidence.

Again, then, just as I learned in the Research Institute, I held back. I started with little suggestions, saying, "Wouldn't it be nice if you tried this?" "Don't you think that could be worthwhile?" and sometimes they would say, "Yes" and sometimes they would say, "No."

So . . . the first case that gave me, in a big jump, a lot of trust in the office was a case that the City Planning Office and the Harbor authorities had not been able to solve for some years—and we solved it in about six months.

The City Planning Office had wanted to take as much of the seashore and harbor area as possible for future urban development. When they said "future urban development" . . . they included public spaces, housing—things of that sort, not harbor activities. And they wanted to do that by just taking land from the harbor. The Harbor authorities knew that, and they objected strongly to it, and clear up until about '85, they had a strong legal right in that the Harbor Authority was governed by state law and state organization. But around '85 that law was changed, so there was a mix of state and municipal responsibility, and I used that as leverage to get into it.

First of all I realized that the Director of the City Planning Office and the Director of the Harbor Authority could not stand each other. It was a personal matter. They had been fighting for years.

So I said, "I'd like the conflict to be put down one level. I think we should start with the senior engineer at the Harbor Authority and me." We would be equal in status, and we would just start working together, to see if we could come up with something.

And they said, "Well, the best of luck to you! That senior engineer is *impossible!* He's the most stubborn defender of the Harbor Authority that you can think of."

I said, "Fine, I'm also very stubborn about city planning."

So I went to meet him, and he was very cold, and really stone-faced, and he started to tell me about all the good things about the harbor. And I found I could surprise him because I could tell him something about the San Francisco harbor, saying "Wow—the San Francisco harbor is really fine. It is professional, harbor-wise, and open in certain places for the public, and that really gives the Harbor Authority support from the public also."

Now this was a sort of a twisted truth, but it had portions to it: you had Fisherman's Wharf, which had a kind of double role. And he had been there, and he said, "Yeah, that's true."

And then as we talked about these professional issues, I said, "It'd be nice if I could have a cup of coffee," and he said, "Sure, you'll have a cup of coffee." And he got a cup of coffee and even a cake in, and then we started to talk about our professional backgrounds, and I said, "I hear that you're actually a civil engineer, that you started out with bridges and things; I did the same."

So we started talking about that, and he said, "Oh, yeah, but . . ." that he was also a military man, and he believed in order and discipline and so on. I said, "Fine, I also went to the military school. I'm an officer, too."

And he said, "You're an *officer,* and you're working in the City Planning Office?"

He had the kind of view that the planners were all these wishy-washy people that would sneak around them and do all sorts of immoral things in their discussions with the Harbor Authority. This was his image of the planners: that you couldn't trust those people; they would always try to fool you.

But he also realized that the planners had good professional training. For instance, he said, "You know, we don't have a man down here that can make a good drawing of the harbor, and of what we want to do in the harbor. But you up there, you have several people who can make nice sketches and fool people—and show that you can build nice housing there at the harbor. But they won't show a nice ship!"

I said, "Sure they can do that, if you ask them to."

"Do you think," he said, "they can make a drawing for me that makes the harbor look nice?"

"Sure."

He said, "I bet you can't."

I said, "At the next meeting I'll bring you a drawing."

So we started building trust. We had five or six meetings like that, half professional, half personal, and I did bring him maps and sketches, showing him that the harbor was nice—showing that the planning staff could produce drawings that showed the harbor nicely.

So furthermore I said, "If you'd be willing to discuss the harbor with me, piece by piece, to see what sort of possibilities we have, I can assure you that you will get, free of charge, the help of my drawing staff. But actually, I think you *have* enough money to buy those services from private companies—but I'll start off by giving you the help, free of charge."

He was very pleased. He said that would be fine—as long as he was the one that would control the drawing, so that in the end he would say, "This is OK; I go for this drawing and not that drawing"—in terms of passing it along to the media or the politicians.

I said "OK, we can do that."

So then we started out asking, "How much of the harbor do you *really* need?"

And he would start saying, "*All* of it."

I said, "I don't believe you. That's a bluff. I can see without being a harbor engineer that you don't need it."

This was a time of transition in the technology in the harbor in which you went from one technology which needs a lot of space and many cranes and things of that sort to fewer cranes, and you take in containers instead of bulk. This you could just see by the visual eye without being an expert.

He said, "OK, I'll tell you what. If you will come with me on a Nordic tour comparing other harbors with ours, talking to Harbor Authorities in other ones, we can continue discussing."

So I said, "Yes, of course I'll come with you," and we had a one week tour. We went to Sweden, to Denmark, to Finland, and we saw harbors there and discussed things with the Harbor Authorities.

And then, for instance, when we came to Finland, to Helsinki, which had a harbor about the same size of Oslo, I could tell him, "Look! They have opened up the harbor; they have given away parts of the harbor, and they've got a nice housing scheme there. There's no threat to the harbor—the harbor can still work fine." Because in Finland they did have a little bit more harbor land than we had in Oslo, so it was easier for them to do that—I have to admit that. But still, they could even continue doing that. And he could also see that it wasn't a threat, it wasn't a conflict, if it was done the right way.

He had suggested the tour, but he was surprised by what he found. But also, I have to admit, I had to change my views on some things. I had thought the Oslo harbor was rather inefficient. After the Nordic tour, I had to change my view on that. It was rather *efficient* in fact.

And also I came to realize that the harbor authorities in Oslo had a problem that not many of the other harbor authorities had, and that was that they were lacking land. They were not lacking shoreline area, but they were lacking land behind the shorelines, land to store things, land to shuffle things, land to transfer from sea to vehicles or to rail—compared to the other ones, which meant that they had to pay attention to how they used the land.

Well anyway, within six months we developed a kind of negotiated process between us, and that's a public participation issue. Actually, the end came in two steps. The first step consisted of two documents—a map and a series of unresolved issues that the two directors could not agree on. That was an interesting development in itself, since presenting these unresolved issues built a sort of trust between the parties. But the politicians thought that the different stands on these issues should be resolved. They demanded that the negotiative process be continued, and they put pressure on the leadership to find solutions.

The second and final step ended with a plan that we could agree upon, except for two small issues. And then I opted for something that had never been done in Oslo. We gave the politicians two alternatives: one which the City Planning Office endorsed, and a second one that was different on two small issues that the Harbor Authority wanted. And the politicians gave the Harbor Authority their support, and that was important, because then *we* gained even *more* trust—that we could present things in our way for the politicians and the harbor authorities would gain.

So this plan settled the border lines of what should be future urban development areas and what should remain harbor land. They all had to give. We both had to give and take. The Planning Director wanted much more land than we were able to give him. But he agreed that, "OK, for, say, the first 15, 20, 25 years, that's OK. Then we can come back to it." And the Harbor Authority also—they actually wanted to keep more land.

But instead of fighting about it for ten years, there was at least now some agreement to go ahead and work and then renegotiate in the future. A harbor action plan was then developed, parallel to the plan in which the land use was determined, which we called a partial municipal plan.

So this built confidence with the Director of the Planning Office. He said, "Well, if you've been able to get this settlement in six months—something we haven't been able to solve in ten years, then there must be something in the technique you're using."

Now, to be fair, I should add that we did have an excellent basis for detailed negotiations in materials we had from a large architectural competition which was held in the early 1980s regarding urban development along the waterfront. We also used a parallel competition between two private consultants to create alternative solutions for our discussions.

So all that, in a way, gave me the trust to design a different model for future planning cases. It gave me more confidence also, but of course it gave it most of all to my own professional staff, because now they had seen that sort of technique work.

Remember, they were not used to this. For instance, when they did urban renewal, and they talked about public participation, it was in the more old fashioned way. You go out with a sketch and say, "Look: This is what I think is good for you." And some will not be able to understand the sketch at all, and they would think, "Well, what should I comment upon? What should we do? I won't say anything."

And some will say, "This portion is really good; but *this* portion we don't think is good at all." And the planners would say, "Why do you think so?" And the people would say, maybe, "We're lacking trees" or "There's not enough place for the kids." And the planner would go back, and he would say, "Well, I think they still could use the space for the kids over there," or the planner might change the plan and then go back again.

But it's not really a negotiated process at all. You listen to something, and you decide what you will hear and not hear, and what you will do and not do. When you've done that a couple of times, then you say, "Well, I've done participation. Now, here's a plan as a result of that process." And I don't think I'm exaggerating. That was about the way it was done. So I wanted to do it differently.

In the full profile (Forester 1994a), Rolf Jensen continues to describe an extensive participatory planning process for the Oslo waterfront, a process involving home owners and environmentalists, commercial and recreational boaters, the press and politicians. But from his account so far, we can learn a good deal about issues of deliberation in planning and design processes.

We would be mistaken to dismiss this story as being about a simple two-party negotiation, for these two were representatives of others, and their hardly private discussions encompassed broad public issues. Furthermore, his story makes clear that while resources of available land were exchanged, both parties also learned and came to reformulate issues along the way. What seemed initially to be a zero-sum impasse, with both sides stuck, antagonistic, and intransigent for years, became a vastly improved working relationship with increased confidence, respect, and regard for the other along the way. What began as a stalled negotiation, with the directors of the two agencies not being able to stand each other, became a successful public deliberation involving engineering expertise no less than design renderings, consultations with independent professionals in other cities no less than mandates from the city council.

Notice that the difficulties of this public deliberation involved far more than the hunger for land on the parts of two city agencies. Jensen makes clear that the whole professional tradition of planning (and engineering too perhaps) seemed to militate against such deliberative processes. The top-down, detached expert model of the planner doing analysis and issuing recommendations to politicians hardly suggested, supported, or encouraged Jensen's work.

But if the planning profession's reading of their mandate and status was an issue, so was the particular political, institutional history of these two agencies. For ten years this issue had simmered, with the agency directors' own feuding relationship not helping in the least.

Jensen suggested that the dispute be "moved down one level," so that the staff might explore solutions free of the tension between the top officials, but he had been warned about the engineer, "Well, the best of luck to you! That senior engineer is *impossible!* He's the most stubborn defender of the Harbor Authority that you can think of."

This story is about institutional pressures and the professionals' taken-for-granted ideologies or their operative planning theories (top-down or deliberative?) no less than it is about the political psychology of these two professionals, the drama of the political interaction between these two adversaries who managed to become collaborative partners.

Jensen seems to have brought a subtle, powerful, and fresh planning theory to his work, one that combined respect for the profession of planning with that for the complexities and uncertainties at hand. Jensen echoes a central, if still underworked, theme in planning practice: the need to learn, the recognition of the unknown that can be known, the keen political and practical sense of knowing that you do not know and that you need to learn, to listen and to develop the informed good judgment that comes from appreciating the perspectives, insights, and needs of others. If an incrementalism had been the antidote to overly top-down planning, it led directly, Jensen knew, to processes of dialogue and negotiation:

"How else," he had said, "can you then do the stepwise incremental work of planning, and how could you otherwise cope with pluralistic wishes? You cannot cater to all of those, you don't even know what they are; you have to look for them in some way. And how do you look for them? By getting into dialogue with people—and as you get into dialogue, you also enter into negotiations."

His point here is less to recommend a conservative incrementalism, deal making, than to stress the need for planners to listen and learn, and in doing so their need to pay more attention to those with whom they are working—more attention of a particular sort, not more presumption of their needs, but more humility. Yet Jensen stresses that this work requires no "downgrading" of the professional calling of the planner, no giving up of commitments to equity, to public welfare, to the goals of a progressive planning agency. This work requires, he tells us, "downgrading the image of your own profession, having more professional humility—reflecting," in his words, "first of all upon what you're doing and being more open for more negotiated approaches."

Having this professional humility and downgrading "the image of your own profession" seems important here for at least two reasons. Certainly arrogance on the part of planners and designers will not help them learn effectively from other citizens. But Jensen's point here seems particularly directed as well to the planners' political sense of themselves: averse to "theory" but nonetheless captive to a particular top-down theory, these planners' "image of their own profession" had subverted their practical abilities to work with others, to learn from them, to negotiate well and collaboratively—as Jensen and his staff in this case, though, were happily able to do.

Jensen's story of thawing out Mr. Iceman of the Harbor Authority obviously does not imply that successful planners should be military officers or that they should visit San Francisco's Fisherman's Wharf. It does suggest that planning deliberations involve exploring and building working relationships as much as they involve research and analysis, expertise, and visual presentation skills. Jensen's story is sketchy but still instructive.

Think only of the perceptions, the images, of the planning staff that the Harbor Engineer seems to have had, as he had inherited ten years of impasse with these planners. He doubted that the planners could be trusted: they seemed to be sneaks, saying one thing in their discussions and then telling the politicians something else. Worse than that, they seemed to take advantage of their aesthetic skills by drawing sketches that falsely made housing schemes look good, while never, of course, making the harbor attractive—doing *justice* to the harbor.

Here we should consider how professional training and education can build walls rather than bridges between people, how a civil engineer might distrust an architectural sensibility, how an "analytic" type might distrust "softer, social sciency" planners. We see such walls separating academic departments on campus, so we should not be surprised when they complicate political and professional life in our cities.

Jensen suggests further that deliberation in planning depends substantially on planners' abilities to listen—not only to know that they need to listen, but to be able to do it, and to pay attention sensitively and responsively too. As he makes clear, sometimes this will mean listening with one's feet: together this city planner and harbor engineer visited several other harbors to expand their common sense of possibilities and constraints; looking and listening together, they learned about new options together, just as they learned about one another.

As Jensen put it, "He had suggested the tour, but he was surprised by what he found. But also, I have to admit, I had to change my views on some things. I had thought the Oslo harbor was rather inefficient. After the Nordic tour, I had to change my view on that. It was rather *efficient* in fact. And also I came to realize that the harbor authorities in Oslo had a problem that not many of the other harbor authorities had . . . they were lacking land."[10]

Jensen learned here too not only about harbors and efficiencies, but about what was important, what mattered. His sense of the facts changed, but his sense of value, of priorities and real needs, changed too.[11]

This story does not imply that successful deliberations depend on our taking week-long tours of Scandinavian port cities any more than on having a staff of architects who can produce pleasing drawings. The central lesson of this story lies elsewhere: in the *transformation of adversarial expectations to collaborative exploration,* of inherited and defensive distrust to small exploratory steps, "piece by piece, to see what sort of possibilities we have," of macho self-righteousness to the willingness to learn practically, politically, and professionally.[12]

Even if no cookbook recipes can painlessly transform political relationships in such ways, these deliberative possibilities are nevertheless real, practically and politically significant, if not very well understood in the planning, public management, and design professions. Too often the rhetoric of "interest-based bargaining," "participatory action research," or even "reflective practice" can obscure rather than clarify possibilities for deliberative and transformative learning. We need not only labels, pithy prescriptions, and general rules (helpful as those can be), but both a better understanding and better images, even literary images, of how these transitions encouraging and promoting public-serving planning deliberations are really possible (Nussbaum 1990a).[13]

Now consider two shorter accounts of practice from planners in the warmer climates of Mediterranean Tel Aviv and Haifa.

Architect-Planner David Best: "It's Just a Sketch!"

David Best is a British-trained architect and planner who has practiced in Israel since the 1950s. After recounting a major planning controversy over the location and layout of an extensive new housing development in East Talpiot, on the southern side of Jerusalem, Best spoke directly to the day-to-day challenges of deliberation, listening, and learning in the planning and design process (Forester, Fischler, and Shmueli 1997):

It often happens that clients don't really know what they want, but you've got to be very careful to be sure that they really don't know. Very often an architect can assume that the fellow doesn't know, but in fact he knows better than you. Sometimes, although the client doesn't present it in the right way, he knows. So you've got to make the effort to find out really what he wants. He doesn't always say what he wants, or he knows what he wants, but doesn't know how to tell you. This happens quite a lot.

Very often a client comes to you and tells you things which seem to be wrong, but what he tells you is not really what he wants. They have a problem of explaining what the issue is, and what the problems are. They're simply not trained to be able to transform a need into directives—and there you must be very careful. I believe there's one very important thing: You've got to give people a hell of a lot of credit at the beginning, and you've got to try to find out, from them, really what they want. You mustn't come, saying, "Oh, no—it's not what you want. This is what you need."

To find out, you need time. This is the critical thing, time. You need time to be able to think a thing out. There's always a question a client asks you, "I want a timetable." My technique, when giving a timetable, is always to put plenty of time at the beginning. I'll tell him, "I don't mind doing the production drawings very fast. But I don't want to produce a sketch in a week."

He says, "I want a sketch; it's only a sketch, it can't take a long time. I want a sketch!"

I tell him, "Look here, the sketch is the most important thing. Translating a sketch into a set of working drawings can be done quickly," and now with the computers, it's even quicker. But in the sketch, there's not just the general layout—the sketch should contain the critical details of the concept because it's the code, the "DNA," which is important. The very smallest issue can be critical and can influence the overall concept, and discovering the real code is vital—and it can be a detail, or a decision about materials, or it can be a decision about a certain structural system, or even a decision that you should go to a certain consultant and get information from him, because that information can be essential to developing your general idea. That needs time. I call it "time to think," and it's one of the most important elements of the planning process.

You've got to build up a good knowledge of how the client thinks, of what his problems are, what the issues are. When you've got that, you've got the confidence to talk to him and to press him further."

"I'll give you an example," he continued. "Two weeks ago, some people came to me, the representatives of a neighborhood association, and said they wanted to build a commercial center of about 750 square meters in their neighborhood. It should look good, and it should be quick, and it should be cheap—a little "shopping mall."

They said, "Can we have a sketch in about a week?"

I said, "No. I want to think about it for a couple of weeks, and then let's have another talk." And I went around, and I saw what's going on. Although I've designed commercial centers, I hadn't done any for about three years, and there's quite a lot of new things being done now. And I came to think, "They don't need a mall, certainly not a miniature mall. What they want is a little group of buildings—a cluster, around an open space, something like a little piazza, possibly a playground. But there's problems, because if I create a little piazza, there's going to be a place where the kids will hang out and make a bit of noise, and this site is near existing houses. So that cluster should be done in such a way that the entrance and the opening should not be towards the houses, but towards the valley. Within two weeks, that is, I had a good idea of what sort of thing they wanted, and it was different than what they had conceived. . . ."

I had discussed the project with my office staff. I'd insisted on getting an aerial photograph, detailed survey drawings of the topography and the road network, and I began to make the first sketches. I reached a point where I had a preliminary concept.

Now, a preliminary concept doesn't mean a final concept, and I don't show clients preliminary concepts, I speak about them. Because if you show them something before you yourself are convinced, you can have the greatest problem of explaining afterwards that that is *not* the solution.

But I was able to tell them, "Look here. One solution is a mall, but there are other solutions as well, and other solutions could have a freer layout of buildings. And I'd like to investigate this, so I'll show you both."

This is a very important point because clients want to see alternatives. Although in actual fact there's a danger too, because providing alternatives without criteria for evaluation is really just letting them make the decision. So you must virtually come to some sort of decision yourself, but you must be ready to be convinced that it's not right. But you cannot give them alternatives without giving them the explanation of which is better and which is less good. You can't just say, "Well, here're the three alternatives, take your pick!" Because then you're relinquishing your job as being an architect, in my opinion.

So I will give them alternatives. I'll say, "Here, first of all, since you've come with an idea, I want to show you what your idea looks like—and I'd like to show you other things as well. In my opinion, I think you should discuss and consider very carefully these other alternatives, because there are certain reasons for having done them. If you make this little mall and so forth, you're putting everybody into some sort of a straightjacket—everybody has got to pay for the air-conditioning, and everybody is together in the management, and that may not be so good, because this is a little neighborhood, and if you have a small cafe, the owner will want his own sort of territory, and want to put his tables out in a certain way that is not going to interfere with someone else. There're a whole range of issues here, as well as aesthetic issues relating to building in an area where a lot of little houses create the urban scale.

David Best, architect and planner, is telling us many things about deliberation in the planning process. First, he tells us about language and ambiguity, the need to look past the literal, to be able not to get stuck on, "Well, what you said was . . ." We need to listen, as Best says, "to find out really what the other party wants. He doesn't always say what he wants, or he knows what he wants, but doesn't know how to tell you."

Second, he tells us about the dangers of professional presumption: "Very often an architect can assume that the fellow doesn't know, but in fact he knows better than you. . . . You mustn't come, saying, 'Oh, no—it's not what you want. This is what you need.' "

Third, he tells us to expect confusion, not only because the issues are complex, because other parties and clients will often have different backgrounds and education, different languages, but also because the other parties' proposals are likely to be internally conflicting, partial and initial proposals rather than fully worked out ideas.

Fourth, in the context of this ambiguity, confusion, and professional-client relationship of power, he provides what we might take to be the first rule of public deliberation. He tells us, "You must be very careful. I believe there's one very important thing: You've got to give people a hell of a lot of credit at the beginning, and you've got to try to find out, from them, really what they want."

Best is not telling us, in telling us to be "very careful," about problems of hearing. He is hearing every word the client says, and hearing the words is not what is worrying him, not what he is telling us to be careful about. He is asking us to be careful not to presume more than we must, to "give people a hell of a lot of credit" rather than assuming we know what they really want amid their confusion. "Giving them credit" means more than giving them respect; it suggests appreciating when they have considered seriously their circumstances and needs. "Giving them credit" echoes the wonderful passage in Robert Coles's *The Call of Stories,* in which he warns professionals to resist the sneaking temptation of the "rush to interpretation" (1989:14).

Best tells us, fifth, about other pressures on public deliberation too. Not only do professional presumptions operate (we can think here, too, of presumptions about race, gender, and class), but the other parties themselves will urge that "rush to interpretation," the rush to premature design decisions and a narrowing of the pie they are to negotiate: "He says, 'I want a sketch; it's only a sketch, it can't take a long time. I want a sketch!' I tell him, 'Look here, the sketch is the most important thing.' "

So Best tells us not just about the passage of time and "time to think," but (sixth) about structuring the professional protection of that time to think, the professional design of the design process to allow that time to think, explore options and alternatives, recast the problem, and re-view the issues.

But Best tells us more than that too. His time is not spent as a professional recluse, for part of it is in service of getting to know the client as well as his or her "problem." Best's "time to think" is also a time to get to know the other, to gain understanding and, more important, seventh, to build confidence, to lay the groundwork for future discussion, evaluation of alternatives, and decision making. He says, "You've got to build up a good knowledge of how the client thinks, of what his problems are, what the issues are. When you've got that, you've got the confidence to talk to him [echoing Rolf Jensen here] and to press him further." "To press him further," he says, always leaving purpose, negotiation, and power in the picture.

Perhaps as important here is what Best tells us about the creation and discussion of alternatives: "Providing alternatives without criteria for evaluation is really just letting them make the decision. So you must virtually come to some sort of decision yourself, but you must be ready to be convinced that it's not right. But you cannot give them alternatives without giving them the explanation of which is better and which is less good. You can't just say, 'Well, here're the three alternatives, take your pick!' Because then you're relinquishing your job as being an architect."

Here he tells us a great deal about the demands of public deliberation and the contributions that design professionals can make by providing design options, criteria of evaluation, ways of appreciating and valuing pros and cons, strengths and weaknesses of options that trained eyes may be able to show the less trained.[14] He argues that design professionals cannot simply throw down the options and opt out: they need to give arguments and come to tentative decisions with "the explanation of which is better and which is less good," and they must realize their own limits, be open to counterargument, "be ready to be convinced that it's not right."

This is a striking example of Jensen's "professional humility," and what Hoch characterized as a deliberative "pragmatism": providing criteria to enhance perception and understanding, providing arguments to recommend options, and remaining open to counterargument, to being shown that you are all wet. This is a picture of a professional architect-planner no less committed to ends than to his or her clients, no less able to work collaboratively than adversarily. This is a picture of a planner-designer able in part to counteract the pressures of professional ideology, bias, uncertainty, ambiguity, and political pressures to rush to judgment—pressures threatening to reduce more public deliberation to manipulation or strategic bargaining.

Transportation Planner Baruch Hirschberg: Deliberation and "Diplomatic Recognition"

Baruch Hirschberg, the Haifa-based transportation planner mentioned at the beginning of this chapter, helps us to understand still better what David Best called that "one very important thing": giving "people a hell of a lot of credit at the beginning."

Hirschberg had told me that sometimes there was "a lot of emotion" in local transportation planning cases, and I wanted to follow that up.[15] He contrasted his meetings with two other parties, one working for the country's most powerful environmental organization, the other for a powerful public agency. Of the first Hirschberg said,

> He's an environmentalist who knows how to talk to planners. With him we can talk for hours. He'll come, and he has a very fertile mind; he keeps coming up with, you know, "What about this?" "Did you try this?" "Look at that . . ." "Let me know . . ." and he gives us wonderful ideas. And I like that, it's mutual. . . .
>
> I learn something from him and he learns something from me. He knows the territory of northern Israel better than I know this office. He will just tell you everything. "No, you can't have the tunnel there because the geology isn't right. It's soft chalk, and it won't work."

I don't know how he knows. He knows everything. So you talk, and you play with paper and pencil, and you draw a line here and you draw a line there, and you come up with new ideas.

Then Hirschberg describes the other:

On the other hand, the head of [the agency], he just sits there at this meeting, for instance: "I have the legal right to prevent any new roads entering this land, so just take this off the bulletin board, it's not going to be. Forget it."

Now I get very annoyed with that. I just told him, "You know, I don't have to come all the way from Haifa to listen to this. If you have no intention of participating in a constructive discussion, then there's no point in meeting with you. Because if you want to veto, sit in your own office and veto. We're trying to look for solutions to problems."

I wanted to learn more about what Hirschberg would have to say about how he and other planners might handle these kinds of adversarial encounters. How, I asked, did he work with local residents or agency staff when they were suspicious, angry, distrusting? His answer cut to the heart of the politics of planning processes—and of democratic politics more generally.

He began simply enough, saying first, "I try to explain what I'm trying to do and to listen to the problems of the other side, whoever it happens to be."

But I wondered: "Listen to their problems?"—fine, but what did that really mean?

He became more specific:

Seeing if I can do something to help them with it. There isn't always a solution. But people need to perceive you as attentive. I think maybe the word I'm looking for is "legitimacy." If you give legitimacy to the needs of your opponent, even if you can't answer them, even if you can't come up with a solution, then they act differently towards you.

I won't mention names. One of my colleagues is a very, very good road planner. But the environmentalists say, "When you sit with him, you just see it in his eyes that we're an obstacle, we're just something to be overcome. We just see it in his eyes." Then they won't make any compromises with him because they feel that he's not there to compromise with them. He's just there to somehow overcome them and get on with his work. Because, basically, he doesn't give them what I call, "*diplomatic recognition*"—the recognition that they, and the interests that they represent, are just as legitimate as the interest that I represent or that you represent.

There isn't always a solution that takes all those interests perfectly into account, and there are times when one interest will have to be subordinated to another, but you have to make the other person feel that you are aware, that just as important as it is to you to have a really good road going up to Kiriat Shemona, it's just as important to him to preserve very special geological configurations like this which are unique there. Once you get that basic message across, then you can talk to people. That basic message is this "diplomatic recognition": that, "You're talking about something that's valid. It's a *real problem*."

If it's not there, what happens is basically there's no communication. One side will say: "This road is absolutely necessary, and this is the place: there isn't any other place." The other side says, "There is no possible way in the world in which you're going to put a road here." Then you just get people digging in.

What I do is this: I always try to find, in every organization—and in my experience it's often possible to find—somebody you can talk to. The question is to find the person, to find the person and to hope that he (or she) has enough influence and power in the organization so that you can actually do business with them. It's not always possible, but it's usually possible. So then there's someone who will listen to your problems, and you listen to their problems and try and find some sort of synthesis to the thing. It isn't always possible, but it usually is.

I wondered how much Hirschberg was claiming here. I found that although he had no great expectations about being able to change people, he was still clear that planning could mean learning together, practically:

But to actually educate people—it's nice if you can do it. But it's not always possible. I don't consider that's what I'm here for, in a sense that my job is to take all these hard-bitten people and make them more malleable. . . . I don't have a degree in psychiatry, and I don't really want to deal with that.

But to the extent that I can, I try to . . . get people who are immediately hostile—for various reasons, and having nothing to do with me personally—to the point where they're just, well, willing to talk to me. And usually, the best way of doing that is just by listening to them.

But what did he mean "just by listening"?

If you listen to them, and they see that you're taking seriously what you're being told, then they'll listen to you. It has much more to do with human nature than with planning or anything else. It's just making people feel that you take seriously what they have to say to you.

Soon after that, I asked him finally, what he considered to be the major lessons of his own practice, and he returned to his sense of recognition:

I would say that the main thing, really, is that when you give other people "diplomatic recognition," even as a tactic, it changes them. It's . . . like in international politics: the handshake between Rabin and Arafat has a significance far beyond the fact that Arafat and Rabin shook hands. It changes them. Neither of them is going to be the same for that handshake.

I think it's the same thing in planning. When you give other people "diplomatic recognition" professionally, even as a tactic to soften them up, *you* end up changing too. Because then you, having recognized them, have to take them more seriously. You can't pretend: it's something that's hard to pretend about. You can't say, "Oh, yes, everything you're talking about is wonderful, and it's very important, and so on . . . because if you don't really believe it, and you don't act upon what you're saying, then you'll be found out, you know.

In these striking reflections Hirschberg echoes the sentiments and practice of Hoch's Chicago planners, Oslo's Rolf Jensen, and Tel Aviv's David Best as well.[16] But he adds depth to their accounts too.

He gives us a glimpse of the all-too-rare case of planning and design as creative, exploratory, even playful deliberation when he describes his meetings with the environmental organization's planner, whom he describes as both knowledgeable and questioning, probing, wondering: "He has a very fertile mind; he keeps coming up with, you know, 'What about this?' 'Did you try this?' 'Look at that . . .' 'Let me know . . .' and he gives us wonderful ideas."

As he echoes Rolf Jensen's and David Best's allusions to the power of the sketch, the power of visual inquiry, Hirschberg suggests what we might aspire to in these deliberative moments: "So you talk, and you *play* with paper and pencil, and you draw a line here, and you draw a line there, and you come up with new ideas."

But Hirschberg certainly does not imply that to be the typical case. In the tougher cases, where suspicion, distrust, and anger are more likely to color his working meetings, he tells us that what seems basic to human nature, "just listening," is really quite complex moral and political work (cf. Rorty 1988, Forester 1989, 1998b, 1999b). He speaks of recognizing the legitimacy of the others' problems: "If you give legitimacy to the needs of your opponent, even if you can't answer them, even if you can't come up with a solution, then they act differently towards you."

Hirschberg's point here is a practical one—neither a general ethical recommendation about the virtues of respect and the dignity of the other, nor a general epistemological point about relativism, as if every claim was equally sound (cf. Putnam 1990). Without such recognition, he seems to tell us, the likelihood of finding solutions together, combining deliberative and adversarial behavior, is not great. Notice that as he calls our attention to relationships as well as to the substantive issues at hand, transport efficiency and environmental protection, he is recommending a particular kind of engaged, attentive, acknowledging relationship be developed, practically and professionally, rather than the one of his colleague who seems to see others merely as obstacles to be rolled over Menkel-Meadows 1995, Gilligan 1982, Tronto 1987).

As Hirschberg explains what he calls "diplomatic recognition," granting the "legitimacy" or, better, the "seriousness" of the other's problems, he clearly does not mean that such recognition implies agreement, that granting such "legitimacy" means believing that the other's claims are always true, or that their demands are justified. He does not argue against evidence; he does not suggest weakening one's resolve; he does not back off from substantive commitments. He wants instead to probe, to ask "What about this?" "Did you try this?" "Could we . . . ?" to explore and come up with new ideas, to imagine and construct the future in new ways.

But he does warn us: we need to convey to others that we can appreciate that their problems are real and matter, that they face issues about which they care, as we care about ours. He warns that unless we convey such recognition that they are serious players, that we take them seriously whether or not we agree with them, we are likely to be left with only strategic bargaining, if not intransigence, where otherwise we might have been able together to create new options, to gain together from more deliberative discussions, from even modest forms of joint problem solving.

Hirschberg is teaching us more about deliberation, too, for he argues that not only do the deliberating parties learn about one another and their concerns as they recognize one another, but they change in the process. As he suggests, he makes this point not as a therapist but as a practical planner encouraging public learning and change. He sees in his work that as recognition is given and enacted, not just intended, by parties, they can create new, more deliberative working relationships, a new basis for going on practically together.

In their differing contexts to be sure, Rolf Jensen, David Best, and Baruch Hirschberg each shows us challenges and possibilities of fostering practical public deliberations in a messy political world. Each of them warns of the dangers of professional and institutional power and presumption damaging the planning and design process. Each of them teaches us about the difficult work of recognizing "the other," taking them seriously, giving them "a hell of a lot of credit," and slowly, cautiously, and deliberately creating the basis to work together.

These practitioners teach us, too, about the complexities of listening and learning in the face of contentious histories, uncertain futures, fluid relationships, and agency rivalries. Each one teaches us about the interweaving of technical and political judgment, about the possible contributions of professional expertise to public learning. Each spoke of the complementarity of visual and political imagination: drawings of the harbor matter, but so does control. "The sketch" matters, and so do time to think, time to explore, time to learn (Goldschmidt 1991).

Each of them teaches us about the real possibilities of exercising a practical public imagination: shaping harbors, neighborhoods, and regions. In the politically astute work of these practitioners and the planners and designers like them, public deliberation means city building in practice.

Jensen and Best and Hirschberg, finally, teach us that *deliberative encounters are there to be created* within conflictual institutions, between organizations and actors who face uncertain consequences and ambiguous mandates, between citizens, politicians, developers, and other planners with agendas of their own. This lesson, also at the heart of Hoch's study, *What Planners Do,* suggests that the widespread concern in planning schools with dispute-resolution processes is right, but perhaps not right enough. We need to understand negotiations and participatory processes not just as they may move from zero-sum to mutual-gain outcomes, but from these narrower notions of bargaining and exchange, now, to more politically and morally sophisticated, but no less practical, notions of public, democratic deliberation.[17]

III

Deliberative Practice Creates Public Value

5

Beyond Dialogue to Transformative Learning: How Deliberative Rituals Encourage Political Judgment in Community Planning Processes

[Even when they want to,] people don't always say what they really mean.
—Anonymous

This chapter explores the ways we learn about value in deliberative settings.[1] In planning and many other kinds of participatory processes, such learning occurs not just through arguments, not just through the reframing of ideas, not just through the critique of expert knowledge, but through transformations of relationships and responsibilities, of networks and competence, of collective memory and memberships.

The argument here assumes neither that planning is participatory nor that it is democratic, just that deliberative conversations about value—about the interpretation and aptness of goals and means—are inescapable aspects of practical action.

This chapter argues in effect that many analyses of dialogue and democratic argument do not go nearly far enough to do justice to the learning that dialogical and argumentative processes can really promote. Inspired by liberal models of voice and empowerment, many analyses unwittingly reduce empowerment to "being heard" and learning to considering seriously local as well as expert knowledge. Participation is thus reduced to speaking, and learning is reduced to knowing—and the transformations of done-to into doers, spectators and victims into activists, fragmented groups into renewed bodies, old resignation into new beginnings, are lost from our view.

Much more is at stake in dialogic and argumentative processes than claims about what is or is not true (as crucial and essential as factual analyses of health risks, for example, certainly *are*). At stake too are issues of political membership and identity, memory and hope, confidence and competence, appreciation and respect, acknowledgment and the ability to act together. The transformations at stake are those not only of knowledge or of class structure, but of people more or less able to act practically together to better their lives, people we might call citizens.

In this chapter we look first at four accounts of participatory action research to portray the richness of deliberative settings that centrally involve issues of learning. These cases range from health and housing problems in rural Venezuela to community development work in New York City and East St. Louis, from economic development efforts in rural Norway to feminist work in an urban shelter for battered women in the United States.

Then we turn to the difficulties of analysis. Here—to hint at the punch line—we find that participatory rituals provide participants not only with dialogue and argument, but with more of relevance than they anticipate, with more of value than they at first appreciate, with possible relations with others they could not foresee, and so with a literally surprising, deliberative political rationality far richer than accounts of decision-making rationality or rational choice allow.

Urban and Rural, North and South Case Experiences

Learning in Action Research in Venezuela

Cano Muerto is an agricultural community of twenty-two households in the Estado Cojedes, Venezuela.[2] Among the participants in this case of participatory learning in 1986 were the *campesinos* (community members), a sociological university-based group (sponsored by the World Health Organization), the Ministry of Health, the governor of the state, and the ruling political party, Accion Democratica. This account closely follows that of Silverio Gonzalez, a participant from the university team (personal correspondence, 1993):

The initial issue was defined by international agencies in Venezuela, particularly the Panamerican Health Organization (PAHO), which contacted the university group because it had demonstrated results in sociological fieldwork, even though those results had not been related to health issues. The issue, at the beginning, was how to deal with the control of Chagas's disease in the tightening economic conditions of the Venezuelan state.

There were no more resources to continue very costly control programs like modern housing construction for campesinos. Since Chagas's disease is a parasitic disease related to poor housing conditions—the triatomine bug finds favorable conditions to colonize the house in roofs of palm and walls without plaster—they wanted an answer from a sociological view.

In the initial traditional research, with deep interviews of the household heads of more than three hundred families, in forty-two small villages, the campesinos' culture showed the research team that the issue had to be reformulated. The problem for them was the house, not the disease. They wanted to have new houses or to improve the houses they already had, but they lacked resources.

The research team proposed to change the issue from bug control (by building unaffordable houses of cement and steel) to house improvement by supporting materially and symbolically the campesinos' knowledge of housing construction of mud and wattle.

In a more participatory approach, the team proposed to try the idea in one rural community, Cano Muerto. They received political and financial support from the governor of the state and, with some reluctance, from the ruling party.

With the community, the team agreed to focus on housing improvements of the campesinos' own houses, with the campesinos doing the work or finding those who would do it. Campesinos and the project team (the university team backed by the governor) negotiated, family by family, the design of improvements—what, when, and how improvements would be made.

The project team committed to provide credit (twenty years, without interest) in terms of material resources dedicated to improve housing conditions. Materials were to be distributed according to improvement progress in the house. The university group agreed to deliver materials for the improvements complete and on time and to assign benefits without political party manipulation.

There was no initial consensus on the improvement program among the twenty-two campesino families. In the beginning, even those who agreed with the idea were not very enthusiastic participants. They did not trust the governor or these strange people from the capital's university. The situation changed radically when the project team complied with its first commitment to bring materials for the house improvements.

"In our first encounters," Silverio Gonzalez wrote, "the campesinos only said, 'Yes,' but they meant, 'Let's see.' Then after the first and second material distribution (in response to their action), they became more talkative, more strictly responsive to the team and more responsible to their commitments to the program."

Gonzalez reported:

> The results surprised positively everybody. In six months, campesinos improved their houses according to their culture and with lower cost to the state. At the end of this pilot improvement program, the university had to retire from the town and campesinos continued their improvement activities on a broad scale, struggling with public agencies to get public services for the town. They got roads, water, electric power.
>
> The town changed a lot in one year. Houses in bad conditions for more than ten years have been improved—in five or six months—for their inhabitants, using their own technology and with little material support from the outside.
>
> Everybody (campesinos, regional government, university, health authorities) wanted that positive action, but nobody believed it was possible. "Verbal agreements" are abundant in Venezuelan public policy, but when there were actions according to those agreements, there came an explosion of sense and energy among participants.

When the mud houses were improved, there were no more bugs in them; however, these good conditions are ensured for only two or three years. The houses need regular maintenance (closing cracks in the walls, avoiding palm roofs, etc.) to ward off the bugs.

The governor got political support and developed a wide housing policy based on this experience. And the Ministry of Health, with pressure from PAHO and WHO, organized a national program based on that experience. But the ruling political party did not appreciate losing control over the situation where the university changed the rules of the patron-client political system. The new broad policy was quickly oriented to their interests, and campesinos' expectations of the program were manipulated.

So how do we see "learning" and even "learning about value" here? The health experts learned about a more effective way to control Chagas's disease than to build new houses. University and agency staff members and political officials as well seem to have learned about the capacity of the campesinos to act for themselves. The campesinos learned that this was not going to be another case of broken promises; when the materials were delivered, they became enthusiastic and effective participants.

But more too, the campesinos were then able to make—they learned to press effectively—further demands on government to procure successfully "roads, water, electric power." Not least of all, the governor and health agency developed new policy measures to extend this experience. The ruling political party learned something too: here was an opportunity for manipulation of credit and benefits.

But how does such learning take place? Through "dialogue"? Through the discovery of information? Through a practical teaching by example? Through a critique of expert knowledge? Through trial and error testing of options? We will return to these questions below. Consider now a case of participatory action research set not in rural Venezuela but in the urban United States.

Learning in Participatory Action Research in New York City and East St. Louis

On the Lower East Side of Manhattan, university students and community members worked to plan for and protect a large and well-known community market threatened with closure. In East St. Louis, community members worked with a student-faculty team from the University of Illinois, Urbana-Champaign, to do neighborhood planning and increase public funds serving community needs. As recounted by Ken Reardon, the lead faculty member involved in both projects, these cases vividly suggest the rich learning possible in action research processes (Forester, Pitt, and Welsh 1993, Reardon 1993, Gaventa and Horton 1981).

Consider first Reardon's brief account of the market project:

> We have to do things like the experiential learners do. You go in with an orienting theory of how that community works and what the problem looks like and why. You then gain much more experience in the community with these people you're collaborating with by going out and searching for a lot of data, information that will either confirm that view of the world in those relationships or confuse it. Reflecting upon that information generates new assumptions, new theories, new hypotheses. But you don't just stop there: you take that knowledge and try to

apply it to solve some immediate problems. In other words, you actively experiment with this knowledge you think you have acquired to see if it has the desired impact. It's this cycle that goes on.

No one goes in as a blank slate, with no notion of what they think they're doing there: what problems they're facing, what the independent variables are, and what the intermediate variables are and how they're going to impact those. But they get in there and they go through this very rigorous effort to collect a lot of additional information about these problems—and all of a sudden it begins to change in its complexion.

Referring to a trend in which larger supermarkets like Wegman's or Shoprite have been threatening to put smaller competitors out of business, Reardon continued,

It's like the example of the street market project we did. What we really thought we were studying was a supermarket, when in reality we were studying a social institution that was about the preservation of certain traditional cultural values and an institution for sharing those understandings of different communities. That was what was unique about that institution. The fact that they sold cheap food was incidental to it. That's a wonderful example of our orienting theory: we were going to create Shoprite; we were going in to Wegmanize this street market.

What's the perfect visit to Wegmans? You're in and out in fifteen minutes and you don't talk to anybody. You can park your car right in front of the door. That's it.

We were thinking about design, social organization, and systems to deliver Wegmans, because that's what we thought was valued by the people who went there. When we actually began sharing what we thought we were learning—about how it would work and how it would need to change in order to be more like Wegmans—we realized that wasn't at all the way people understood the importance of that institution. We were studying the wrong species of institution. And that's unbelievable when you think about it. We collected the surveys, we had all the statistics, we could report levels of significance. But it just didn't mean a goddam thing to the people we were working with.

We weren't even collecting the right information for the kind of place it was. We had this theory that we were studying a mass volume supermarket in which the real factor was price, since it was an economic institution. In reality it was about a whole other set of values and experiences.

Reardon explained what happened:

We did have instruments which had both closed and open-ended questions. The fact that there were closed-ended questions in the instruments shut people down so that they gave closed-answer-like responses even to the open-ended questions. So we never got the kind of information that would've really helped us to realize that we had a model for fixing this place that was based upon the wrong phenomena.

We moved too quickly into designing the instruments on that project after just meeting with the clients once or twice. We did not spend any time doing participant observation at the site, and we didn't do much informal interviewing. Now we tend to do a lot more schmoozing and hanging around and informal discussion with people in the very beginning to get a better sense of things. We then try to test what we think we're seeing and looking at with the sponsoring committee, which we didn't do as much before. I think we would now know, earlier on, if we were in the wrong department at the zoo, studying the wrong species or wrong phenomenon. But I'm sure we'll make errors in the future as well.

Turning to the East St. Louis project, Reardon pointed to additional issues concerning organizing, participation, and leadership:

The objectives are not only to develop a set of programs and initiatives that address basic concerns, but to do it in a participatory fashion so that folks can continue doing that kind of creative problem solving on their own. The outcomes we're looking for are not only improved physical conditions, but also an increased ability on the part of the local community-based organization to do planning and programming and an increase in the number and quality of community leaders in a position to facilitate this process. The process is as important as the program outcomes it produces. There's a real commitment to taking the time to fully involve not only the local leaders who at the beginning are the officers of the group, but also a growing number of members of the organization. We use the planning process to recruit new long-term residents of the community and neighborhood who have never been involved in civic activities and get them involved in the process through the planning effort.

The role of the university research team was always problematic, it seems:

We are trying on a continuing basis both to increase the number of folks who are involved in sitting down and working with the process and to begin to shift more and more of the responsibility for leading it from the researcher in the form of the University faculty member to both active students and neighborhood residents who're engaged and involved in the process. It's actually transferring skills and power through the process of doing the neighborhood plan.

Developing the capacity to provide leadership . . . is one of the critical criteria one would use in evaluating the effectiveness of action researchers. It's not just that they've come up with some interesting solutions to three or four immediate problems that you'd consider to determine whether the action researcher's done a good job. The question is also: what's the ability of current residents and leaders to continue the process and to train other people in their own community?

Reardon explains how this process worked in East St. Louis:

Doing a common work plan that lays out different roles, we try to encourage the residents to take as much responsibility as possible. I wouldn't really say it's assigning people because it's a voluntary effort and none of us like to be assigned, but it is a process of laying out steps that need to be taken and specific tasks that would accomplish each step, and then trying to encourage people to do what they can . . . to contribute to moving the process forward. Part of what the action researcher attempts to do is, in a developmental way, to try to move people through playing very small public roles to taking larger and larger roles based upon having successfully carried out increasingly complex organizational and planning tasks. . . .

As you begin to go through the planning process, you try to identify a series of levels at which people can get involved and hopefully move up to the point of increasing their skills and their confidence and their sense of ownership over the process.

Somebody comes, they get excited, and they want to help. The next thing that they would do is the sign-in at the meetings. Then in several months they might do a minor report as part of the agenda. Later, they might have a major speaking role, presenting something that needs to be discussed by the whole group and acted on. Maybe at nine months they can actually be chairing a sub-committee of the neighborhood group or presenting part of their program and why it's so important to somebody in power. So it's a developmental process.

That process may transform the organization itself:

As time goes on, you develop a core of people and they can also help with [follow-up]. An independent leadership group begins to develop, and then there's always a public review of progress that's made on the project at each meeting so that people know that they are going to be held accountable. People are chosen for leadership positions in this kind of a planning effort because they've carried out work on behalf of the group in a consistent manner, and you try to help people appreciate that. This cannot be an organization that selects their leaders based upon who makes the most eloquent speech alone.

Reardon described the evolution from a dubious skepticism to active leadership qualities:

Part of it is working through that initial skepticism people have. Once people begin getting involved in doing some new things that they haven't done before, they usually like it. They then see themselves getting recognized in the planning process as a result of doing the work, and that becomes very important to them. They had gone out and surveyed ten or fifteen people on their block before as part of the Democratic party or something like that, but then nobody said, "I think it's very important that you present the information at the next meeting." They never had a public leadership role: They did the work but they never got recognized.

The fact that they are now getting recognized really drives it home to them because you're talking about people who have never had a chance to articulate their concerns to people in power. As the organization begins to build, develop, and grow, it has the ability to offer these people that opportunity that's very important to all of us: to be heard.

This type of support is really giving them power, empowering them, and exciting them, and it's paying off for the community. You can begin to see people change in the process: they are more likely to voice their concerns. If they are not listened to, they pursue the point and feel like they have a right to do so, whereas before they might not have. They have learned not to back down, to feel like they have a right to negotiate with people in power over important things and have a right to participate in the process. For many people, that is an incredible leap in their consciousness and sense of confidence in what they deserve and who they are. They carry themselves differently.

These two excerpts suggest that participatory action research projects involve many forms of learning. Community members and researchers alike may collaborate to generate research questions, approaches, and new understandings. In the New York market case, the university researchers came to realize early on that they had collected the wrong information to learn about the wrong problem. They had presumed the market was an economic institution with problems of inefficient operations; only in consultation with market members did they come to realize that other problems were more pressing.

In addition, Reardon points out, there's a "developmental process" of building leadership in participatory action research (PAR) processes. Residents become involved, become recognized by others, and recognize too that they themselves can act in ways they had not before. Here the residents are learning from acting together as they bring issues before public bodies, organize neighborhood attention to issues, lead meetings—and so develop skills, confidence, and ownership in the planning and organizing process.

Dialogue is important to such learning, as we will see. But more is learned than is said, and if we overemphasize the talk and the dialogue in participatory action research, we risk missing what is truly transformative about such work. Before showing how and why that is so, we consider two more illuminating cases, one involving economic development in rural Norway and the other a battered women's shelter in a large urban area in the United States.

Learning in Action Research in Rural Norway: Public-Private Economic Development Strategies

Consider now Morten Levin's account of an action research project concerned with rural economic development in Norway (Levin 1993). Levin first describes the three phases of their work as originally intended:

> The initial thinking was to create an action research (AR) process where key people in the community could interact and collectively use their creativity and skills to create a network of business and public activity based on local resources. . . . Key members of three local communities accepted this approach to economic development.
>
> The first phase of the project centered on creating new businesses through designed, managed cooperative activities. Since action research is a generative process, learning from phase one created questions for later development, action and research. The second phase aimed at building a network among critical constituents of the three municipalities in the region. . . . A development conference and follow up actions that involved key people in the region was used for this purpose. The challenge was to overcome existing intermunicipal conflicts by creating a network of professional leaders who shared a common purpose. The third phase focused on developing relationships between private and public sector organizations to support business development. The aim was to search for and develop a new and effective role for the public sector in supporting business development. (p. 194)

Levin notes that the action research process itself might have promoted network development:

> The aim . . . was to encourage cooperation among key parties in the region to stimulate and support network development. The development strategy rested on the assumption that "Industrial networks are not designed by any single actor according to a master plan or a strategic decision. They emerge and develop as a consequence of exchange between semi-autonomous, interdependent actors." . . .
>
> Action research also provides a process for inventing new organizational forms shaped by participants' needs, intentions, and decisions and for making adjustments in these forms over time based on learning from the AR process. . . . AR methodology was chosen because it required close interaction of researchers and actors in the field. Research aimed to bring about a mutual learning process that would result in creating networks that supported economic development. (p. 196)[3]

Levin explains that the researchers effectively reframed the project at hand:

> The original project concept focused on exploring business opportunities inherent in new information technology. The concept rested on an "outside-in" assump-

tion of development: begin with the requirements of technology external to the community and match internal resources to these. This . . . was seen as the starting point for attracting new organizations to the region. The basic goal . . . was to encourage large national companies to set up subsidiaries in the region.

But the outside researchers questioned whether the proposed "outside-in" development concept would work. They also disagreed with the centralizing values inherent in the concept. Consequently, they reframed the project to focus on indigenous development with local participation and control using the AR process. As reframed, the project constituted an "inside-out" development process that used AR to build ways of discovering meaningful economic development activities. . . . [In a steering committee meeting with mayors of all three communities present along with local politicians and county development agency administrators,] the first item covered a review of individual projects in the existing development plans in the communities. Although the review process took a long time, listening to the mayors' descriptions and reflections provided all participants with a rich introduction to the problems and possibilities in the region. The process also gave researchers insights into the key community members' models of economic development. . . . [Subsequently,] a "search conference" was the first method used to create an arena for social exchange. (p. 201)

The search conference technique, Levin explains, combined information generating with what Reardon might have called a developmental social process:

The search conference resulted in a new social structure because it brought together different sets of people organized to carry out a vaguely defined task through action task-forces. The new informal structures that emerged from establishing and supporting taskforce activities over time served as a bridge that connected developing a vision and planning work at a "search conference" with existing organizational structures in the community. . . .

In Hitra, where the new mayor was strongly supportive even without having been a participant at the earlier conference, the second search conference attracted strong local attention. It seemed that participation either conferred or acknowledged social status in the community. The general climate of the conference was very enthusiastic. After the conference, one of the participants in line with others, stated, "I know you have to force us to go through this ritualistic process of searching, but the results and the engagement that the conference ends up with are very exciting." . . .

Search conferences created a forum for participants to meet and to engage in cooperative efforts. Everyday life in small municipalities does not grant legitimacy for such close cooperation. . . . Designing a process of discovery and triggering an active push for involvement is required to bring about [effective public-private cooperation]. . . . Creating legitimate fora for mutual exploration, development, planning, and action seems to have allowed the parties to begin to engage in worthwhile joint activity. (pp. 204, 208, 211)

In addition to the information processed, Levin points to a series of secondary outcomes:

> The formation of new organizations, positive unintended outcomes grow[ing] from failures or partial failures . . . and . . . the emergence of a public-private tourism network [in which] cooperation among tourist industry organizations has increased [and] a new role for the public sector [developed, in which] municipal administration and the tourist industry jointly developed a cooperative effort based on a common initiative. An area of shared understanding and commitment emerged from the search conference procedure, and this common understanding, in turn, shaped the new, mutually dependent cooperative network organization. . . .
>
> The search conference appears to have been a powerful tool for merging interests of various constituents and for producing practical results. Search conferences provided a neutral setting and provided a structure for social interaction on topics of general interest [that] led individuals to participate around superordinate goals. Working toward shared goals, in turn, tended to reduce conflict among representatives of various groups who participated in the search process." (pp. 211–213)

Levin concludes,

> A network is a system that results from the social interaction among representatives of member organizations. The AR process shows potential for developing and shaping networks through the use of several interventions. The planned change process creates new fora in which people interact in new ways to grapple with common issues. Old traditions, conflicts, and expected patterns of behavior are put aside through the developmental process. New expectations emerge through cooperation in both identifying goals and developing ways to reach them. The AR process also creates a means of identifying undiscovered talent already resident in a community. . . . These findings [come from] an action research approach that focuses on co-generative and participative development of ideas. . . .
>
> [But] no network will last forever. It has to be constructed and reconstructed as part of a continual growth process. A challenge for the public sector is to discover and invent key tasks inherent in the role of broker in creating, modifying, and recreating networks. (p. 216)

Levin refers to several forms of learning here—even more than he notes explicitly.[4] He tells us that the "essence of action research is the creation of a new social reality through discourse that encourages and supports learning." We see an emphasis not only on ideas and argument here, but on building new networks, creating new organizational forms, and perhaps most important for the AR process, an emphasis on the creation of fora in which participants can meet to consider the challenges in front of them and the disputes entangling them.

Before turning to the analysis of learning in these settings, we examine another urban case of participatory action research.

Race, Class, and Gender Interwoven in Participatory Action Research

Consider finally an urban example in which issues of race, class, and gender were central. We quote briefly from a profile of a university-based PAR practitioner, Michelle Fine (Forester, Pitt, and Welsh 1993):

It's always been clear to me that research in the absence of action was inadequate, but it became more clear to me—or as clear to me—that action in the absence of inquiry was fundamentally flawed. So when I became a social psychologist, I wanted to develop a way to generate methodologies and epistemologies that were rooted in activism but that had both the capacity for collective reflection and the responsibility of theorizing—which is what I think research brings to activism. . . .

One example of my work is a project a co-worker and I did with a battered women's shelter in a large East Coast city about two years ago. The shelter had contacted me because they were interested in self-analysis; that is, they were interested in an evaluation for their own interior conversations. We agreed that our work would be that of feminist outsiders who shared a sense of the politics of the organization, but that we were interested in pressing a set of questions that the organization by itself didn't feel as though it could surface and carry with enough trust to pull it off only on the inside.

One obvious example of these issues was race and class relations within the staff and between the staff and residents. They felt as though it was useful to have somebody come in, collect the analyses as they were voiced by the residents and the multiple layers of staff, and then create a forum where people could begin to address class and race issues. Another was differential perceptions of "success"—or the goal of the shelter as perceived by staff and residents. For residents, success was leaving a violent home, even if it meant going back tomorrow, but having a place to go to. For staff, it was leaving a violent home, staying at the shelter, getting out of that violent home and never going back.

Those were differences—that they were aware of—that were only emerging as tensions. What they wanted was a way to document and to feed back those issues in order to further a conversation they knew they needed, but they didn't feel as though there was enough trust for it to happen.

Fine recounts the process:

The directors of the center were very aware of this, but they felt as though if they initiated the conversation, it would feel very top-down. They were interested in documenters who shared a set of politics with the shelter, were committed to collecting information to support transformation, and then who would document the work.

We explained our process to every level of folks in the organization: the Board, the residents, the multiple levels of staff. We explained that we saw our job as collecting these stories and feeding them back to a mixed group so that people could hear each other's stories. We understood that we were trading on our race and class and outsider status in order for everybody to get a hearing. That is, people were more likely to listen to us tell the stories of the residents than the residents themselves. And yet, everybody agreed that it was important to provoke those conversations and then set up mechanisms on the inside so that those conversations could happen organically rather than be artificially initiated by outsiders.

We, mostly my co-worker, collected all these stories. We wrote up a document that reflected each of the perspectives. Each group got a copy of their section so they could review it and comment on it. In some cases, we changed what we wrote; in other cases, we just added their responses because we felt as though we had to have some integrity around the authority of the text, so it wasn't just feeding back quotes. We had to stand for something; we had some analysis in there.

We then convened a session where people could talk about the document, and we created some ground rules—such as people shouldn't be saying whether or not somebody else's perspective is true or not, but instead how they hear it, how they feel about it, what it means for the organization. It was out of that exercise that the Board, the staff, and the residents developed a process for, firstly, having some task forces that dealt with specific internal issues that needed ongoing attention, like racism and classism, and, secondly, they created a set of retreats for their own process so that residents and Board and staff would always be in conversation about the nature of the shelter.

One of the issues that they were addressing as a policy issue, including all the constituents, was whether to go broad or deep: whether the shelter should handle 2,000 women a year, all of whom can stay for a week, or whether to really focus on the needs of 90 women and their kids over a sustained period of time. For the first time, that was a decision that was not simply going to be made by the Board, but it was going to be made by the Board in consultation with decision-makers among the staff and decision-makers from the resident group.

Here again we see issues of trust and the creation of space in which participants' stories can be told and heard. Here we see more clearly than in other cases the influences of race and class as they may divide staff and clients, researchers and others too. We see learning at several levels: about the claims of appropriate goals for the organization, the needs of the women served, and the very structure and capacity of the shelter organization itself to deal with its internal conflicts. Again, dialogue is essential, the space for the sharing of stories is essential, but learning seems to take place in ways that go beyond the increasingly clear understanding of what someone else has said.

Each of these cases can teach us about the social structuring of deliberative discussions. In each, participants come together to discuss means and ends at the same time. They meet and refine their ideas of common possibilities, the problems that confront them, and the options they have to respond. If we can clarify the modes of learning described here, we could then ask practically: What organizational or social structural designs can enhance or obstruct, support or hinder, these forms of deliberative learning?

On the Design of Deliberative Political Rituals

Two Powerful Models That Help, But Don't Help Enough: Beyond Understanding and Dialogue to Practical Transformation

We learn from more than arguments and voice in participatory settings, but how we do so is far from clear. In negotiations, participatory groups, and ordinary meetings too, we learn not just with our ears but with our eyes, not just with our heads but with our hearts. We come not only to hear new information we find relevant, but we come to see new issues that need our attention. We come not only to revise our sense of strategies, but to develop new relationships with others too.

The public participation and participatory action research literatures draw often on two powerful and related models of learning, both dialogical and argumentative, one indebted to John Dewey, Chris Argyris, and Don Schön, the other to Paulo Freire and Jürgen Habermas (Argyris and Schön 1978, Brown and Tandon 1983, Dewey 1930, Freire 1970, Gaventa 1991, Greenwood 1991, Habermas 1975). Both are important, but both are too limited; we need to refine and supplement these views.

The Deweyan model focuses on the ways we learn in dialogical action together by testing our hunches, assumptions, and suggestions of action. By making a move based on an initial strategy, being surprised at the consequences, whether positive or negative, and reframing our strategy as a result, we learn as we "reflect-in-action" (Schön 1983). As Reardon (1993a) put it, "Reflecting upon that information generates new assumptions, new theories, new hypotheses. But you don't just stop there: you take that knowledge and try to apply it to solve some immediate problems. In other words, you actively experiment with this knowledge you think you have acquired to see if it has the desired impact. It's this cycle that goes on."

The Freirean model focuses on the ways we learn in dialogue by probing our political possibilities of speaking and acting together: Who owns the land, controls knowledge, and might yet have more control over their lives? Whose definitions of problems and solutions, of expertise and status, of power and powerlessness perpetuate relations of dependency and hopelessness?

The brilliance of the Deweyan model as refined by Chris Argyris and Don Schön lies in its reflective pragmatism, in its ability to make sense of the trial-and-error reflection-in-action of practical experience. The brilliance of the Freirean model lies in its insights into relationships of knowledge and power, voice and growth.

These insightful analyses have been influential and instructive, but they ignore important problems too. Unless we supplement them both with powerful accounts of deliberative practice, preserving what they offer but recognizing a good deal more, we will fail to promote the learning that participatory research, and deliberative democratic inquiry more generally, can make possible.[5] If we do not understand the fuller promise of participatory processes, we will be likely to shape the deliberative occasions of community meetings, workshops, retreats, mediated negotiations, and participatory research efforts in needlessly restrictive ways (Horton and Freire 1990).

This chapter accordingly proposes a third approach: a transformative theory of social learning that explores not only how our arguments change in dialogues and negotiations but how we change as well. At the heart of this transformative account is a view of citizens' performances in safe rituals of participation in which citizens not only pursue interests strategically and display themselves expressively, but reproduce and reconstitute their social and political relationships with one another too. To understand participatory and deliberative encounters as social and political ritual performances means coming to see these as organized forms of presenting and exploring value rather than as going through meaningless motions, as forms thus connecting past memory and obligation to future strategy and possibility, and far more. Far from being empty containers in which dialogue takes place, these deliberative rituals are laboratories, if not cauldrons, of political judgment.[6] But how does all this work?

In speaking together in more or less dialogical or argumentative settings, not only do we learn from *what* others say or do about what they claim, but we learn still more from the way they do it. From the reasons they give, we learn about what others want or believe. But from the way they talk and act, from their style, we learn about who they are, "what they are like," what sort of "character" someone has (or is). In the Cano Muerto case, the campesinos' agreement to government promises was a cautious, "Let's see." Only when the project team lived up to its promises, showing in deed what they claimed in word, did the campesinos learn that new possibilities were really in the works. So, too, Michelle Fine speaks of the importance of trust in her work with the battered women's shelter, of sharing commitments and a feminist politics.

From the ways that others act, participants learn not only something about who these others are, whether they are arrogant or not, trustworthy or not, reliable or not, but about how they recognize, appreciate, and honor (or dishonor) value in the world we share with them. Checking with members of the market, Reardon's student team found it had misunderstood the entire institution. What mattered to the community members was not so much price but culture, not so much the economic marketplace as the social meetingplace. Listening carefully to the clients and staff of the shelter, Fine and her coworker discovered an important difference in their perceptions of "success," of the basic goals or values the shelter sought to achieve. Curiously, though, participants learn about one another and about significant issues from more than the strategic arguments they hear. As we shall see, they learn not only from arguments but from incidental and quite revealing details, not only from the presentation of survey results but from the surprises that their *participatory rituals* make possible. We can think of participatory rituals as encounters that enable participants to develop more familiar relationships or to learn about one another before solving the problems they face—for example, the informal drink before negotiations; the meals during focused workshops; the small break-out groups complementing plenary problem-solving sessions; the early story-telling phases of mediation processes, and so on. Participatory rituals are encounters in which "meeting those people" comes first, even if it serves the secondary objective of "solving our problem." On such occasions we discover that we learn about our problems through, and as we learn about, other participants too. Pushing for solutions too soon—before affected parties have been able to listen to one another—can end up taking far more time than preliminary participatory rituals might.

The Importance of Messiness: Letting the Details Surprise and Teach Us

In many participatory settings, a good deal is fluid and unclear. Issues are formulated and reformulated. Relevant history is debated. Senses of future possibilities vary. Information is not perfect. Understanding is not guaranteed. Even if the words are clear, there is still far more to learn than meets the ear. A Norwegian planning consultant of much experience expressed this in a simply and elegantly understated way when he said that in participatory processes, even when they want to, "people don't always say what they really mean."[7] In that one line lie enormous implications for democratic practice, for critical listening, for the work required to appreciate and respect what people really do mean and value.

We learn not only from the points people make but from the details they present—and often the unintended details. This is particularly important because many times we are at risk of being held hostage to our own presumptions of what others "really want." To respond to others and not just to our presumptions about them, we need to assess the particulars of what they say—as bogus or accurate, as predictable or surprising, as genuine or exaggerated, and so on. Reardon speaks of the power of "additional information" and the process of considering it in changing participants' initial expectations: "No one goes in as a blank slate, with no notion of what they think they're doing there: what problems they're facing, what the independent variables are, and what the intermediate variables are and how they're going to impact those. But they get in there and they go through this very rigorous effort to collect a lot of additional information about these problems—and all of a sudden it begins to change in its complexion."

In many meetings, the basic protocols and ground rules can be pretty straightforward, but the particulars of what is presented as significant are the sites of exploration, criticism, and learning. Each of the particulars, each of the objects of attention, has a place in a past and future story. In Norway, Levin worked with the public and private actors to generate together new possibilities of cooperation, new ideas, and new networks. In Cano Muerto, Gonzalez and his research team found that the problem of health could be approached as a problem of housing improvement and not new housing construction as previous programs had attempted. Investigating the marketplace, Reardon's team discovered a rich history of relationships whose importance was not nostalgic but fundamental. Until the team understood the history of the market, they failed to understand what the market really was, what the market really meant to the people involved. Fine shows us that getting such historical detail can be a political matter, and that participants will only be able to listen to the relevant particulars, to others' stories, in certain processes—in her case, not too "top down" a process.

This leads us to stand the traditional fact-value hierarchy on its head. If value-free facts would be, by definition, without value, really worthless, we can come to see that a claim about a "fact" is simultaneously a claim that something is important, that something matters. So too in participatory settings. The details presented are not mere details, worrisome minutiae (though these exist to be sure), but they are often claims about value, claims about what one party is worried about, wants to gain, is afraid of, wishes to protect, or cares about enough to put on the table for discussion.

The particulars others raise can seem irrelevant at first, and they may turn out to be irrelevant—but they may also turn out to be surprising, suggesting problems or opportunities. Participants may come to see that what *seemed* unimportant is important, what seemed *not* feasible is feasible after all. This "coming to see" (housing conditions as an opportunity for action; tourism as a matter of common concern) is a matter of recognition—quite literally, re-cognition, that is, coming to see the very same thing in a new light, with new significance. From the Cano Muerto case's "surprising" results came new national health efforts, for example. Seen now in a new political light, the particulars of housing improvements valued by the campesinos become not irrelevant but crucial to future possibilities of the government and international health agencies.

Participants learn not only about what others care about, but they themselves learn about concrete details and issues whose evolution or resolution will shape what they can do in the future, what they may hope for (or dread) today, whom they can become tomorrow. In Cano Muerto again, the initial work on housing improvements led to further work and improvements on roads, water, and electricity for the campesinos.

Such learning from detail, and unforeseen detail, is hardly accidental. This learning is organized in a subtle way: not through a simple means to ends, but through varying participatory rituals of meeting, talking, eating, and listening together. So Reardon tells us,

> *When we actually began sharing* what we thought we were learning—about how it would work and how it would need to change in order to be more like Wegmans—we realized that wasn't at all the way people understood the importance of that institution. We were studying the wrong species of institution. And that's unbelievable when you think about it. We collected the surveys, we had all the statistics, we could report levels of significance. But it just didn't mean a goddam thing to the people we were working with. (emphasis added)

In the sharing of concerns came surprise and learning—here and in arguments about the truth or falsity of the survey's results. Promoting such surprise and learning, safe participatory rituals can enhance our own quite limited rationality too.

Letting Stories Supplement Our Limited Rationality: Re-Minding Ourselves via Ritual Performance

PAR participants, like others in decision-making processes, are not all-knowing "rational economic men" with perfect information or perfect self-knowledge. Their predictions are uncertain. Their interests and senses of value are ambiguous. In the flow of activity, they must focus their attention narrowly. But because they are not all-knowing, they can learn and be surprised (Nussbaum 1990:72).

But like other decision makers more generally, PAR participants care about much more than they can focus on at any given moment. They may be discussing job creation strategies or housing needs, and so other concerns not relevant at the moment go unattended: the water quality of the neighborhood, the transportation system they depend on to return home, longer-run goals, and so on. But PAR processes may enable participants to learn not only from arguments about possibilities, but from allusions to what is relevant in the first place, especially when arguments, issues, alternatives, and concerns are multiple, ambiguous, and conflicting.

Decision making, planning, and participatory processes are dances in which the initially relevant can become irrelevant and the apparently irrelevant can become relevant. An action research team, as Gonzalez suggests, may begin a meeting with the belief that housing is the chief concern, but they may soon see that trust in government promises and the experience of past betrayals are even more pressing.

In planning and PAR processes, participants meet and discuss, propose and explore, even as they continue to learn about the implications of their own tentative offers and queries, their own contingent suggestions of "What if we . . .?" and "If we did this, could you do that?" and so on. Not going in with "a blank slate," as Reardon put it, not knowing as university researchers what the campesinos could really do in Venezuela, PAR participants have to do a dance with their own intense directedness, their own singularity of focus, their own danger of concentrating so closely on yesterday's need and strategy that they miss today's and tomorrow's opportunity.[8]

As participants realize the limits of their own knowledge, realize that they have to learn about relevance and significance, that all issues involved with practical options are hardly labeled ahead of time, they can begin to appreciate in new ways aspects of decision making and participatory processes that appear at first to be needless preludes, ritualistic wastes of time, or even distracting preliminaries. Levin quotes a participant who says, poignantly, "I know you have to force us to go through this ritualistic process of searching, but the results and the engagement that the conference ends up with are very exciting."

Participants can come to see that the safe rituals of small group meetings, for example, remind them of concerns they do have—concerns with continuing relationships, needing to deal with each other next week, for example—concerns they may have put out of mind in their urgency to deal with the issues at hand. They may see that the ritualized turn taking of storytelling reminds them of concerns they had forgotten: to take into account Garcia's new enthusiasm about working with Jackson, to be wary of Smith's easy promises, to watch out for the campesinos' being burned once again by campaign promises. They may see, too, in the turn taking, what they "knew" but may have forgotten in their pressing desire to deal with the shelter's issues: that everyone present has a story, a sense of self, varying political resources of support or opposition, of trust or suspicion, of cooperation or resistance (Viggiani 1991).

PAR participants learn—and teach us—that ritualized storytelling or small group brainstorming or group identification of "strengths weaknesses-opportunities-threats" works to *re-mind* them of their own concerns—to bring into new focus values they have, obligations they wish to honor, interests they wish to satisfy even if they *did not have all this in mind at the beginning* of a given meeting.[9] Information begins to take on "a new complexion," as Reardon says, and Fine tells us of the need to structure a continuing conversation through which to explore issues and listen to diverse stories (Garcia 1991, Mathews 1994, Reich 1988).

The official, recorded minutes of previous meetings often fail to accomplish this ritual re-minding process. The words on the page, typically quite abstracted from their context and summarized quite drastically, pale in comparison to participants' actual voice and style as they present themselves and reveal their concerns (or demands).

A simple summary list of issues may never do the job of the initial storytelling, precisely because the summary list is too simple. When issues are complex, when organizational decisions involve or will affect many actors, decision makers (PAR participants) need reminders that will help them identify emergent issues that will matter, particular issues of this new dispute or opportunity and not just more general concerns (cf. Forester 1999b, Cavell 1968). The detailed richness of stories, their seemingly distracting detail, can remind participants of interests and teach them about issues that they may not even have had in mind at the meeting's beginning.[10]

Learning About Value in Ritualized Storytelling Processes

Far more than descriptions of events, stories are forms in which their tellers can offer what they take to be worth passing on.[11] Telling stories, participants present examples of virtue to emulate or vice to avoid: campesinos' control of their own housing improvements to encourage or "outside-in" centralized development strategies to avoid. Telling their stories, participants present promises honored or betrayed, offer insights regarding the real story behind another's facade, or recount a history of who did what to whom, for example, when the governor's last program was implemented. Such storytelling provides not simply a "Who Dunnit?" record, but the quite practical context for future action and judgment.

Because they present what speakers take to be worth telling in the circumstances of addressing the problems at hand, stories are a conventional form in which PAR participants articulate value. In the design of PAR processes, the deliberately ritualized organization of public storytelling, we see the ritualized means through which participants can articulate and explore value together. Levin refers to these as organized fora for the exchange of views. Fine refers to this as furthering a needed conversation about tough issues. Reardon points to the importance even of "schmoozing and hanging around and informal discussion" to learn about what really matters to community members.

Such participatory rituals of telling and listening to stories can work transformatively in at least three ways: to transform identities, agendas, and perceptions of value in the world. Consider each briefly.

Transforming Relationships and Identities First, and perhaps most commonly, conversational or storytelling rituals can produce or reproduce, strengthen or weaken, the public senses of self we call social identities. For example, with the mutual acknowledgments of "Do you . . .?" and "I do," a marriage ceremony publicly transforms (the identities of) two "single" people into (the identity of) a "married couple." Similarly, a weekly religious ritual of participation in "church" can be embraced (or resisted) as it presents its participants, more or less authentically, as practicing members supposedly honoring a religious tradition.[12] Reardon refers to participation in the community planning effort as a "developmental process" in which people come to "carry themselves differently," and to a central dimension of learning in participatory processes that we must not neglect: "Developing the capacity to provide leadership . . . is one of the critical criteria one would use in evaluating the effectiveness of action researchers. It's not just that they've come up with some interesting solutions to three or four immediate problems that you'd consider to determine whether the action researcher's done a good job. The question is also: what's the ability of current residents and leaders to continue the process and to train other people in their own community?"

In participatory processes, the ritualized interactions can shape new identities and relationships. New groups, organizations, or networks, not just arguments, can be created. Participating together to promote economic development or to provide more responsive services, we form new working groups, subcommittees, task forces, and networks. As a result of meeting together, making cautious and contingent agreements together, exchanging tentative promises of "I will do this, if you do that, and then we'll see what else we might do . . ." participants recreate and redefine their relationships, creating at times practical new organizational forms and networks.[13] "Know-what" and "know-how" come to be supplemented here with a particularly social "know-who."

Fine, for example, refers to the shelter board, staff, and residents as developing "a process" for "task forces that dealt with specific internal issues that needed ongoing attention, like racism and classism," and also creating "a set of retreats for their own process so that residents and board and staff would always be in conversation about the nature of the shelter." Reardon speaks of developing not just understandings but "skills," "confidence," and "ownership." So also does Levin summarize the Norwegian action research effort:

> The action research process shows potential for developing and shaping networks through the use of several interventions. The planned change process creates new fora in which people interact in new ways to grapple with common issues. Old traditions, conflicts, and expected patterns of behavior are put aside through the developmental process. New expectations emerge through cooperation in both identifying goals and developing ways to reach them. The action research process also creates a means of identifying undiscovered talent already resident in a community.

Transforming Issues and Agendas Second, rituals pattern attention selectively. Reardon's participatory planning effort focused residents' attention on demographics, physical conditions, and social organization. Levin's search conferences organized attention to the region's economic development strengths and weaknesses, threats and opportunities. Gonzalez and his colleagues refocused attention from costly housing construction to more decentralized and cost-effective housing rehabilitation strategies.

Steven Lukes has written lucidly about the political significance of such ritually organized selective attention. He argues, accordingly, for an approach to the study of ritual that "would go well beyond the conventional study of politics, which, as Edelman puts it, concentrates on who gets what, when and how, on 'how people get the things they want through government,' and focus instead on mechanisms through which politics 'influences what they want, what they fear, what they regard as possible and even who they are' (Edelman 1964:20). . . . It would also examine the ways in which ritual symbolism can provide a source of creativity and improvisation, a counter-cultural and anti-structural force, engendering new social, cultural and political forms, involving what Turner calls 'liminality' and 'communitas' " (Lukes 1975:302; cf. Edelman 1988, Turner 1969, 1974, Kertzer 1989, Moore and Myerhoff 1977, Rosen 1985). We can see rituals, in Lukes's view, then, "not as promoting value integration, but as crucial elements in the 'mobilization of bias' " (p. 305).

In their ritualized forms of collecting stories, brainstorming, recording and reporting small group discussions, creating larger group memory by covering walls with lists of discussed items and topics, PAR participants necessarily focus their own attention selectively. Coming together, they invoke and honor certain values and not others; they read or quote certain texts and not others; they express and respond with certain emotions and not others.[14] They express as a matter of public record certain allegiances and not others. Providing, as Lukes put it, potentially "a source of creativity and improvisation, a counter-cultural and anti-structural force, engendering new social, cultural and political forms," these participatory and deliberative rituals prepare their participants—more or less well—to recognize new issues and attend creatively and responsively to particular struggles at hand.[15]

Transforming Ends: What's at Stake Third, then, ritual performers articulate, organize, and invoke claims of value together.[16] In a search conference devoted to the initial exploration of strengths and weaknesses, threats and opportunities, the ritual participants are asked to step back from a "last-word" analysis to share with one another their informed hunches about what matters, what is worth considering together, positive and negative. Their ritualized stories may often be richer than anyone intends, and so the very occasioning of storytelling allows surprise and learning that extends beyond the initial purpose of any single storyteller.

Inevitably we say more than we intend, and we mean even more practically than we say (Cavell 1969, Levin 1993). Perhaps the most brilliant expression of this point comes in a line of Iris Murdoch, who might help us not only to listen to stories but to look carefully to their sensitive and wise telling—in participatory and deliberative conversations, planning and policymaking conversations, and those of everyday life too. "Where virtue is concerned," Murdoch (1970:31) wrote, "we often apprehend more than we clearly understand and *grow by looking*."

Listening together, we recognize as important not only words but issues, details, relationships, and even people we may have ignored or not appreciated in the past. Listening to tone as well as content, we may recognize not just claims about what the government has or has not done in the past, but a history of betrayal and resulting fear, suspicion, and distrust, which must be acknowledged, respected, and addressed if working relationships are to be built. Recognizing such issues in public, acknowledging them in participatory groups, we come not only to transform our shared senses of what is at stake today in our deliberations but to shape new commitments to new ends and to one another as well. Our learning, our transformation, is both cognitive and collective. In acknowledging another's concern we can come to see it more clearly, test "to see if we have it right," and we can encourage and shape relationships of solidarity and collaboration too. So Gonzalez recounts the developing commitments of "the explosion of sense and energy among participants," and Levin tells us that "new expectations emerge through cooperation in both *identifying goals* and developing ways to reach them" (emphasis added).

But not all stories or rituals are created equal. We have all been there: Smith may drone on and on, not presenting issues of value but boring everyone to tears and driving them to distraction. Epstein's story may in fact be irrelevant or an absurd paranoid fantasy. Jones's story may be obnoxious or distracting or ill timed, all of which may threaten and not serve a participatory decision-making process.

Not any turn taking, sharing of stories, or generating of shared texts will do. If the participants do not have some shared sense of the rules of the game ensuring safety in their meeting together, they may not be able to act together. If they do not have some sense of structure and process, of protocols of turn taking, of appropriate and inappropriate action (and storytelling) in this meeting, they may be too confused or threatened or shy or reticent to participate, sharing what they know, signaling important concerns to others, reminding others, warning others, showing others new options too (Viggiani 1991). If participants do not feel safe and trust each other enough to speak and listen responsively, as Fine suggests, they will hardly work together, and their negotiations and participatory decision making will most likely fail (Pitt 1993).

The Ritual Structuring of Unpredictability as the Ground for Learning, or Decision Making When Interests, Parties, and Priorities Are Changing

Some meetings can be so structured, so predictable, so predetermined that no one learns much of anything new. The designers of the decision-making process may have decided what information shall be explored and how it shall be, and their controlling intentions may make surprise, discovery, and the identification of new issues and opportunities practically impossible. Then again, if someone could fully know ahead of time just what issues and information were really going to be relevant, he or she could not only design wholly predictable decision-making processes, but predict, if not dictate, the precise outcome as well.

But if, more realistically, we *know that we do not know* everything that will be relevant, if we know that we do not know what options we will discover in the process of listening and responding to one another, then we need not so much rigid predictability but instead a structured unpredictability that will help us to ask new questions and consider new answers. This structured unpredictability of ritualized storytelling may be the most important element that we can design to facilitate practical learning by participants about the breadth and depth of their own concerns: it exposes them to relevant but surprising, important but unforeseen, claims (facts and issues, provocations and emotional appeals, and more) that they take to matter.

This unpredictability of the ritual itself—the search conference, for example, or the neighborhood planning process's review of data—enables emergent groups to form a steering committee, a subcommittee, an e-mail group, a network or coalition, and so on. The ritual occasion may structure the safe collaborative possibility of participants' exploring and forming tentatively new roles, new groups, and thus new identities, along with accompanying newly designed norms, rules, agreements, or conventions that articulate how the participants may go on together. The openness of a search conference or a mediation session to a new agreement about roles and rules means that ritual occasions enable innovative self-generation, group redefinition, and innovation that may not be envisioned by any participant at the meeting's onset. Gonzalez suggests as much when he recounts,

> Everybody (campesinos, regional government, university, health authorities) wanted that positive action [of effective housing improvement], but nobody believed it was possible. "Verbal agreements" are abundant in Venezuelan public policy, but when there were actions according to those agreements, [there came] an explosion of sense and energy among participants.

If surprise is essential in the deliberations of actors with limited information and rationality, participatory rituals are the social structures that may stage and enable the unpredictable learning occasioned by such surprise.

From Garbage Cans to Transformative Rituals

James March and his colleagues have described a somewhat analogous process of nonintentional discovery in decision-making settings (March 1988). Painfully aware of participants' limited knowledge and time, shifting preferences and resources, March and his coresearchers chose a somewhat desperate metaphor as they characterized a "garbage can model of decision making." In the garbage can, streams of choices, problems, solutions, and participants meet in unpredictable ways to shape ongoing, complex, and messy organizational outcomes. As March and his colleagues (Cohen, March, and Olsen, 296) put it, "Organizations . . . provide sets of procedures through which participants arrive at an interpretation of what they are doing and what they have done while in the process of doing it. From this point of view, an organization is a collection of choices looking for problems, issues and feelings looking for decision situations in which they might be aired, solutions looking for issues to which they might be the answer, and decision-makers looking for work."

Understood in this way, participatory decision making becomes a good deal richer than even an elaborate bargaining process, as the next chapter will show. Participation itself is assumed to be fluid, with some people paying more attention to some issues than others, having more time at some points than others, and so on. Participants clarify and discover goals rather than take them as given, and so in the process of interacting, they come to value some ends they had not cared for previously. Similarly techniques become not only tools to be used on demand but competences whose presence shapes participants' search for solvable problems and understandings of themselves (cf. Winner 1986). E-mail capacity, for example, will shape participants' thinking about their needs to keep in touch with one another easily; "have computer, will network," the participatory action research saying might go, updating the more common, "Have hammer, look for nails. . . ."

The garbage can model of decision making can be refined or recycled. Choices, problems, participants, and solutions may not so much be dumped into garbage cans, as March and his coauthors suggest, as they meet, interact, and transform one another, in ordinarily structured ritual performances that organize members' attention and shape their emerging exploratory commitments.

Held hostage to its own pungent metaphor, the garbage can model of decision making and deliberation is right, but not right enough. Participants and problems and opportunities do often come together in ill-structured and messy ways, do often find a voice, or fail to, in unpredictable ways, but then more happens too. In their ritualized interactions, participants can come to see one another in new ways, to redefine and reformulate problems, to clarify opportunities, to reorder priorities individually and collectively. The process of deliberation and participation is better seen not only as argumentative or dialogical in terms of who knows what, not only as allocative, in terms of who gets what, but as transformative too, in terms of who comes to create new relationships and act on new commitments in actual practice.

Participants begin deliberative processes with agendas and suspicions, with complex cares, sets of interests, and senses of possibilities—all informed by yesterday's, but not yet informed by today's, experience. By articulating and exploring their ambiguous priorities, evolving relationships, and uncertain options today, participants in ritualized processes of dialogue and deliberation can transform their expectations and obligations as they make new agreements or revise old ones; they can transform their senses of self and other as they take new roles or redefine old ones; they can transform their senses of value and priorities as they come to recognize new issues or reevaluate old ones; they can transform their working relationships as they form new organizations and networks or reshape old ones. Far from finding themselves simply thrown together as chance may have it, as March's garbage can model might suggest, deliberating participants can have a good deal to say about what they do together, about how they listen to one another or fail to, about how they search for new options or fail to, about how they learn and find new ways of going on together, or fail too. They have a good deal to say not just about how they can refine their knowledge of strategies but about how they can transform themselves. Reardon spoke of this possibility in East St. Louis:

> This type of support is really giving them power, empowering them, and exciting them, and it's paying off for the community. You can begin to see people change in the process: they are more likely to voice their concerns. If they are not listened to, they pursue the point and feel like they have a right to do so, whereas before they might not have. They have learned not to back down, to feel like they have a right to negotiate with people in power over important things and have a right to participate in the process. For many people, that is an incredible leap in their consciousness and sense of confidence in what they deserve and who they are. They carry themselves differently.

The process is one not simply of throwing together a group of rational decision makers, but one of changing those decision makers into a more deliberative political body. In this way, the ritual richness of participatory encounters provides the infrastructure, the materials and occasion for a deliberative political rationality, potentially transforming means and ends and self and other, a rationality far richer than typical accounts of optimizing allow. But political power is always an issue here. Gonzalez, for example, tells us both of the campesinos' increased political organization, their successful struggles for roads, water, and electricity, and of resulting manipulation by the ruling party. Fine tells us, too, of

> trading on our race and class and outsider status in order for everybody to get a hearing. That is, people were more likely to listen to us tell the stories of the residents than the residents themselves. And yet, everybody agreed that it was important to provoke those conversations and then set up mechanisms on the inside so that those conversations could happen organically rather than be artificially initiated by outsiders.

The search conferences and workshops and storytelling rituals of action research processes take shape in the light (or darkness) of encompassing power relations. As deliberately diversified small groups meet to discuss problems and opportunities, participants can learn in surprising ways. The same happens in mediated negotiations or joint problem-solving processes: participants find themselves meeting together with supposed adversaries from their own city, adversaries with whom they may have never spoken informally, and they can learn about one another in ways that do not blunt their objectives, do not co-opt them, but make them capable of serving, elaborating, and focusing their interests *more* effectively (Forester and Weiser 1996, Forester 1998a). So Levin tells us,

> Everyday life in small municipalities does not grant legitimacy for such close cooperation. . . . Designing a process of discovery and triggering an active push for involvement is required to bring about [effective public-private cooperation]. . . . Creating legitimate fora for mutual exploration, development, planning, and action seems to have allowed the parties to begin to engage in worthwhile joint activity.

Rather than finding themselves scavenging with March and his colleagues in organizational "garbage" for idiosyncratically interesting objects to work with, PAR participants may find themselves learning in surprising and unpredictable ways as they participate in loosely goal-directed but ritualized performances of sharing stories together, brainstorming possibilities, listing strengths and weaknesses of salient organizations, and so on. These ritual processes enable them to listen and learn without forcing their attention so narrowly that they miss the richness of concerns and capacity that others bring to their encounters. In such ritualized meetings, participants are asked to listen more than to choose, to consider new information and feelings more than to accept or reject them. So Levin writes:

> The search conference appears to have been a powerful tool for merging interests of various constituents and for producing practical results. Search conferences provided a neutral setting and provided a structure for social interaction on topics of general interest [that] led individuals to participate around superordinate goals. Working toward shared goals, in turn, tended to reduce conflict among representatives of various groups who participated in the search process.

So too, reconstituting political space—trying to "create a forum where people could begin to address class and race issues"—Fine tells us that she saw her job as "collecting these stories and feeding them back to a mixed group so that people could hear each other's stories" (Hoch 1994, Innes 1996, Abrams 1991). Mindful of power relations, she and her colleague suggested ground rules for the shelter's discussion too: "People shouldn't be saying whether or not somebody else's perspective is true or not, but instead how they hear it, how they feel about it, what it means for the organization." Fine and her colleague encouraged not so much an informative exchange of arguments here but a transformative ritual of deliberation. In this structured conversation, issues of action ("how they hear it"), emotion ("how they feel about it"), and consequences ("what it means for the organization") might all be explored.

Learning from Structured Complexity: Rituals as Aids to Dialogic and Deliberative Rationality

The ritual richness of participatory processes can re-mind participants of significant issues by bringing before them—enabling them to perceive, to recognize, to re-member, and even to express (re-cite)—items of value they may have not had immediately or clearly in mind. Loosely goal-directed rituals enable them to learn about and acknowledge value, to identify new concerns and issues, without forcing choice.

These ritual encounters allow participants to listen and learn, to watch and see, to feel and identify concerns they were not clear about earlier. These encounters also involve and partially commit participants as they provide occasions for their own performances of storytelling and issue identification. Acting in front of others on such ritual occasions, participants take certain stances of tone and posture and attentiveness. As they tell their stories, their stories tell a good deal about their selves too (Throgmorton 1996, Healey and Hillier 1996).

This much leads to a curious result. If the richness of ritualized storytelling and listening exposes participants to details, to particulars that re-mind them of their own concerns, then the more narrowly they focus their meeting on their immediate purposes, the *less* they may learn in ways they themselves need.[17] Their own strong presumptions about others—lacking better information than they now have—hold them captive, preventing them from exploring issues and options that would help them. If curiosity kills cats, presumption kills negotiators. Their narrow purposiveness can not only blind them, but it can blind others to the interests and values they might probe together, if not fully serve.[18] This, then, is the argument for creativity and play in complex, uncertain, and ambiguous decision-making contexts (Cohen, March, and Olsen 1988).

The richness of ritualized storytelling and listening involves not just any ritual form. The ritual structure provides the vessel but not the message. The stories of participants' concerns can be expressed in a variety of participatory settings, but the more narrowly pre-scribed, expressible in advance (pre-dictable) they are, the less informative, demonstrative, evocative, refreshing, and even redemptive they will be. If participants are to be able to acknowledge one another's concerns, they must be able to address one another. If they are to be able to address their differences, they must be able to listen to one another and to speak responsively as well. Seen in this light, ironically, public hearings are pathological rituals that often minimize responsive interaction and maximize exaggeration and adversarial posturing.

Consequently the danger arises: the thinner the ritualized occasions that might enable the expression and exploration of specific concerns and interests—be these formally structured and time-limited presentations or relatively informal conversations over drinks or meals (or coffee breaks, breakout groups, corridor conversations, or barroom or bathroom chats)—the less will be learned. The challenge here for planners and PAR designers is the structuring of deliberate distraction, the deliberately enabling distraction from participants' own narrowness of focus that prevents them from remembering, recognizing, and learning in ways they wish to.

There is a prisoner's dilemma analogy to be made here—a trap of rational intentions we fall into when our inclination to focus on the sure thing, the interest we find most pressing and clear, drives out *our* competing inclination to learn about more ambiguous, less certain, related ways of satisfying our own interests. Going for the sure thing—the interest at hand—we fail to see that we have still other interests whose mutual satisfaction will serve our interests better still. Precommitment (putting the alarm clock out of our reach so we cannot go for the short-run desirable sure thing) is one response: precommitment not just to *outcomes* (Ulysses binding himself to the mast to resist the Sirens) but by participatory rituals and institutional design to *processes* of listening, consideration, reminder, articulation of memory, exploration of options, delay of premature agreements, involvement of third-party intermediaries, testing of evidence, probing of mutual contingent promises and offers, and so on. So Reardon says, "The process is as important as the program outcomes it produces. There's a real commitment to taking the time to fully involve not only the local leaders who at the beginning are the officers of the group, but also a growing number of members of the organization."

The more purposeful and simple our statement of "positions," to use the formulation of advocates of interest-based bargaining, the more meager our opportunities to learn, to probe new options, and to design strategies that will effectively serve our ends (cf. Fisher and Ury 1983, Neustadt and May 1986:274). Once we understand that we begin deliberative discussions with limited rationality, less information about others, and still less information about the full range of possible agreements we might devise, we can then appreciate that we need not so much bottom-line simplicity in our deliberations but rather a ritually organized, optimum complexity of description and narrative richness that enables us to learn—and that falls far short of chaos and immobilizing confusion.

Such ritually organized complexity honors the participants' senses of direction but not any single purpose; it enables the presentation and articulation of clusters of concerns and interests rather than narrower positions; it allows the consideration of relevant questions before narrowing participants' attention to any overriding one. In this light, the ambiguity of interests and priorities, even of law and political mandates, is an opportunity for participants to explore and work to resolve together rather than only a source of indecipherable confusion.

The search conference is one such ritual design that generates both intentional complexity and deliberate distraction. It sketches a broad purpose, a direction of commitment to be explored together; it invites participants to teach one another about relevant history, facts, techniques, and opportunities. In doing so, participants can learn not only about strategies to serve their interests, but also about one another and about interests of their own that were less salient to them at the beginning of the search process (Greenwood and Levin 1999, Susskind, McKearnan, and Thomas-Larmer 1999).

But more happens here too. As the ritualized groupings and conversations occur with conventions and rules of their own (even those of ordinary small talk in a coffee break), participants find themselves not only bound by these rules but enabled by them as well (Forester 1998a, 1999b). Not only are certain forms of formal argument and posturing precluded at times, but other forms of listening and being heard become possible in these new arenas of meetings of parties and, in some cases, previous disputants. So too do seductive information technologies (interactive computing, multimedia systems, and sophisticated mapping programs, for example) pose both threats and opportunities, for different forms of representation enable different forms of interaction and deliberation, different forms of emotional responsiveness, memory, insight, and perception (Peattie 1987, Nussbaum 1986, 1990a).

Looking narrowly and abstractly at the tough choices before them, PAR participants might make choices (and treat each other) more callously than they must, if they let themselves ignore important values because they cannot serve them at the moment. But here participatory rituals that ask all participants to step back before going forward, to look broadly before focusing narrowly, can help. Ritually structured re-minders of the values at stake in their actions can help to keep participants honest, help to keep them to the standards they respect, help more practically to ensure that their limited attention does not blind them and others to the necessary "costs" or "losses" or "downsides" of their own decisions.

Acknowledging Others: Encouraging a Politically Deliberative Community

Decision making involves not only cognitive choice but social expression, social articulation that divisively or redemptively addresses those served and those disserved, winners and losers alike. So may election winners, for example, often acknowledge in acceptance speeches the virtues of their defeated opponents and recognize the presence and significance of the opposition. So too can the explanation of a political decision acknowledge and credit opposing arguments, not ridicule or ignore them, even as it justifies and seeks to motivate and gain support for another course of action.

Such public and articulated acknowledgment of conflicting and pressing values does not solve a problem; it works ritualistically to re-build relationships and prepare the social basis for future practical action. It works to reassure losing interests of their standing; it works to transform its audience from a collection of winners and losers on one particular issue to a more integral and enduring political community of members respecting one another's differences (White 1985, Susskind and Cruickshank 1987, Susskind and Field 1996, Forester 1998a, 1999b).

Such ritualized acknowledgment of multiple and conflicting values might counteract even a coming decision's potential divisiveness not only by re-minding its audience of the significant concerns of opponents but by encouraging all parties to re-member their common commitments to a political life together. So Fine's challenge is help the women's shelter not to avoid choice but to maintain the solidarity that will allow staff and clients to make difficult choices and continue together afterward. Reardon tells us too of the importance of participants' being recognized, of their initial skepticism giving way to confidence:

> The fact that they are now getting recognized really drives it home to them because you're talking about people who have never had a chance to articulate their concerns to people in power. As the organization begins to build, develop, and grow, it has the ability to offer these people that opportunity that's very important to all of us: to be heard.

Counterbalancing their participants' own limited attention to their pressing purposes, then, rich rituals that allow a rehearsal of concerns, commitments, and values can help their participants to learn in ways they are unable to intend, foresee, or predict. In choosing this or that ritual of storytelling together, participants choose a more hierarchical or conversational way of learning together, of shaping solidarity in different ways, along class lines or not, lines of expertise or not, and so on. But what they will learn in the conversation cannot be foreseen with much confidence in advance.

Into the ritual occasion of sharing stories and concerns, for example, or sharing lists of strengths and weaknesses, threats and opportunities, come concerns and relationships. With the concerns come particulars and facts that matter, details suggesting issues to be explored. With the relationships come evolving possibilities of understanding, of mutual agreement and contingent promising, of collaborative opportunities, of going on together in unforeseen ways. Participatory and deliberative processes, then, are not simply sites of bargaining and trading; they are occasions on which participants can deliberately, if gingerly, transform their senses of self and opportunity, and their practical relationships too.

Conclusion: The Significance of Ritual in Participatory and Deliberative Settings

The analysis of learning through deliberative, participatory rituals suggests that we learn not only with our ears but with our eyes and hearts. We learn not only from surprising information that leads us to propose new hypothetical lines of action to test, but we learn from style and passion and allusion too. We learn to reframe our predictions and strategies, but we learn to develop new relationships and even senses of ourselves as well. In participatory processes, we not only generate arguments, but we construct networks and new organizational forms as well.

This analysis suggests that participatory groups learn in a great many ways. We must be wary of focusing so much on argumentative learning that we fail to appreciate participants' learning of skills and confidence, appreciation of and respect for others.

This account might help us improve on the intriguing but limited "garbage can" model of decision making in which solutions find problems, choices find preferences, and decision makers find work. Transformative rituals enable the deliberative reconsideration of preferences and choices, of problems and solutions, of decisions and decision makers, by enhancing participants' limited attention, information, and rationality.

An analysis of transformative and ritualistic learning in participatory settings can help us to understand the significance of the very messiness, complexity, detail, and moral entanglements of living stories and dramatic role-playing presentations. Far from being simply diffuse or emotional, these stories are accounts of value and identity, of abiding concern and complexities that are ignored at practical risk.

This account can help us to understand that the selective representation, in speech or writing, storytelling or role playing, of problems and choices and decision options always has political and moral resonance, as it acknowledges or fails to acknowledge the depth of human concerns at stake. Social theorists write of the redemptive, value-acknowledging character of rich narratives because they are concerned about the ways that history or literature may do justice, or fail to do justice, to the lives represented there (Myerhoff 1982, LaCapra 1983, 1994, Nussbaum 1986), as chapter 7 will argue at greater length. Members of participatory groups surely have no less obligation to respect the pain or loss that their actions will cause as they make what too easily get called "tough choices."

This account might also help us to understand the oblique and surprising, nonintentional learning we must protect and even enable ourselves to do in practical situations. We know that we do not know in advance what we will come to value and what we need to learn about others and the complex political environments we face. Knowing that, we need not only to offer and test hypotheses, to move and consider the "backtalk" of the situation as Don Schön argued, but we need to look and listen, to engage others deliberatively and probe joint commitments, to allow ourselves to remember values less of the moment but no less to be honored, and to resist what Robert Coles, quoted in chapter 1, so wonderfully called "the rush to interpretation."

Finally, the analysis of safe rituals of participatory learning helps us to resist the temptations of economistic bottom lines, and so to understand that political deliberation, the participatory attempt to reconsider ends and means together, to learn about what we want and what we can do, is far richer than a process of bargaining and trading across pregiven "interests" that we have in any stable and well-defined rank orderings. This transformative ritual account of participatory or deliberative learning helps us to understand that in participatory processes, we may not only argue to test strategies for maximum gain, but we may make and act on new agreements, transforming ourselves and others in the process. We generate new ideas, but also new organizational forms. We develop better practical hypotheses, but also new capacities of action and cooperation. We may transform ideas and ourselves, in unexpected but valued ways, as well. Knowing that we often do not know enough to act well, we need to learn, to deliberate in practice—together—as we seek to encourage participatory planning processes.

6

The Promise of Activist Mediation in Planning and Public Management

This chapter explores the challenges of a newly emerging field: environmental mediation and public dispute resolution more broadly.[1] Although much of what professional practitioners write about mediation and dispute resolution is hype, and often marketing hype, much of what academics write is far too insensitive to the demands of practice—the demands to do as well as one can in the complex institutional settings at hand.

This chapter attends to questions of public dispute-resolution practice in a relatively novel way, by exploring themes of democratic political theory and republican political theory in particular. By public dispute resolution, I refer to mediated and facilitated processes seeking to resolve disputes concerning resource allocation, public policy formulation, or governmental rule making (including, but extending beyond, environmental mediation, for example). Along the way, the argument suggests how feminist theory and social theory offer important insights. Throughout we assess the practical character of public mediated negotiations and its implications for planning practices. The chapter concludes by sketching a research agenda that would take the tasks of political philosophy seriously and suggesting that public sector mediators ought to be seen not as experts, judges, or rule-bound bureaucrats but as new Aristotelian friends—"critical friends" of the disputants as persons, interested parties, and citizens of a larger public world.

The chapter has five sections. It first sets the stage for what follows by considering the public presentation of self by three prominent suppliers of public dispute-resolution services. It next suggests that planners and public managers as well as mediators and facilitators managing public dispute-resolution efforts face similar practical and political challenges:

to nurture deliberative processes that produce enduring agreements while being inclusive or properly representative and sensitive to power imbalances at the same time. Third, it explores the character of public participation and political deliberation, assessed as the central values of a modern republican politics. The fourth section poses research questions exploring mediators' judgments about power and representation. The last section challenges the adequacy of predominantly empiricist approaches to the study of public dispute resolution; it argues that normative concerns of political philosophy must be taken seriously if we are ever to debate not just the mileages on the map but where it is we wish to go.

Public Dispute Resolution's Public Face

In the public world, disputes are everywhere. Seeking to find a home for a halfway house, a local church may find itself dealing with the resistance of local neighbors: "Not in my backyard" is the familiar cry. A local government may try to respond to pressures from environmentalists ("Protect open space"), real estate developers ("Modern Condominiums! Sauna and Pool!"), and affordable housing advocates ("Housing we can afford!") too. Seeking to improve an airport or a highway link, a regional transportation agency may try to work with municipalities at odds with one another, with a variety of local business groups and interests, and with citizen and environmental organizations. Industry representatives may threaten layoffs if stiffer regulations are passed legislatively or adopted by regulatory agencies. And so on.

Reflecting the variety of such disputes and the difficulties of resolving them, recent literature describes negotiation and mediation efforts in cases that involve, for example, the regulation of environmental resources (land, water, and air), historic preservation, capital improvements and infrastructure development, facility siting, commercial development, housing provision, and growth control (Rivkin 1977, Johnson 1986, Dotson, Godschalk, and Kaufman 1989, Carpenter and Kennedy 1988). Typically in these kinds of disputes, organized parties and affected publics are many, issues are multiple, and law provides boundaries rather than determinate directions for action.

In such cases, practitioners and academics alike have argued that mediated negotiation offers many promises: to produce agreements more quickly between disputants that will go further toward meeting the interests and needs of the affected parties—agreements that will be more stable, agreements that will reflect the virtues of a participatory decision-making process (Bacow and Wheeler 1984, Moore 1986, Susskind and Cruikshank 1987, Carpenter and Kennedy 1988, Ury, Brett, and Goldberg 1988). Yet skeptics worry about imbalances of power and expertise, however "face to face" and "voluntary" the process; the danger, they fear, in having the environmentalist lamb lie down with the corporate lion, is that the lion will be hungry for another lamb in the morning (Amy 1987). Obviously, both these promises and the skepticism they engender need to be investigated.

So while the advocates of mediation typically promise collaboration, protected relationships, and more efficient outcomes, legal analysts in turn worry about deal making that can not only go unauthorized by, but preempt, the review of the courts (Folberg and Taylor 1984 Carpenter and Kennedy 1988, Edwards 1986, Grad 1989). And analyses by philosophers, as they often do, continue to dazzle but frustrate us with nagging questions that seem to have too many, and contradictory, answers. For example, David Luban (1985:402–403) asks us once more how Rich and Poor should negotiate to split $1,000: $100 may produce as much pleasure for the poor woman as $900 may for the rich man; $500 would be equal amounts to each; yet the poor woman may need $950 and the rich man only $50 (to take a cab home). Are these all "good outcomes" to their negotiation (Raiffa 1985, Barry and Rae 1975)? Although the literature on public dispute resolution is growing quickly, its volume still outpaces its quality.

In this section, we seek simply to characterize the general types of claims that public dispute-resolution advocates make by turning to the public information such advocates distribute. Here, in formats running from glossy, full-color booklets to photocopied news clippings, we find the promises made by those in the business of public dispute resolution. To get the flavor of the enterprise and to note similarities of theme, we review three promotional statements briefly.[2]

Consider first the way the nonprofit New England Environmental Mediation Center described its mediation of "site-specific" disputes:[3]

> In many instances problems are effectively and appropriately resolved by a regulatory agency or the courts. However, in other instances, the parties can better solve the problem themselves if they can establish a forum for negotiating with one another.
>
> Mediation is a special kind of negotiation in which an impartial "third party" helps those involved in a dispute to reach a mutually acceptable agreement that resolves the problem. Many professionals—including lawyers, consultants, and regulators—help parties resolve disputes. What makes mediation unique is that the mediator has no stake in the outcome of a case and no power to impose a decision. Mediation creates a situation where participants are free to explore options without the demand to posture and politic before decision-makers and other authorities. Throughout this process, the mediator works for all the parties, seeks to focus attention on issues and substance, and encourages creative problem solving.
>
> . . . During [the] exploratory period, the mediator analyzes which parties should participate, whether the parties can agree on the issues in contention, and whether there is enough uncertainty about the outcome of events, or enough frustration with the current administrative or judicial process, for parties to consider some form of joint settlement effort. . . . *Mediation begins only if and when all parties agree to it.*
>
> . . . Mediation is a safe process . . . with a set of groundrules and safeguards that protect *all* interests.
>
> - All parties must agree to mediation before it can begin.
> - Participation is voluntary and any party is free to withdraw at any time;
> - The mediator has no authority to impose a settlement;
> - Discussions and draft agreements are typically kept confidential;
> - No party waives legal rights or responsibilities because of its participation;
> - No agreement is binding until signed by the parties.
>
> . . . Mediation can yield solutions to environmental and natural resource controversies that are *better than* those which result from purely administrative or judicial proceedings. Often, it reveals unexpected and acceptable ways to solve complicated problems. Parties avoid much of the misunderstanding, delay, expense, and complexity of litigation, prolonged administrative proceedings, or inaction. Participants can often reach solutions which are not possible when working separately, and can do so without compromising their fundamental interests.

To many, this is the familiar portrayal of environmental mediation. The mediator is an "impartial" third party with "no stake in the outcome" and "no power to impose a decision." No decision maker, the mediator focuses "attention on issues and substance" and "encourages creative problem solving." Thus, participants can "explore options" together "without the demand(s) to posture and politic before decision-makers and other authorities." The process is "voluntary," for "any party is free to withdraw at any time."

The process, we are told, complements (and does not displace) judicial or regulatory processes: "No party waives legal rights or responsibilities because of its participation."

Why can the resulting solutions be "better than those which result from purely administrative or judicial proceedings"? Because, this presentation claims, the parties can learn: they may find "unexpected and acceptable ways to solve complicated problems." Working together, participants can "reach solutions which are not possible when working separately." Such collaboration allows learning and joint settlement, which may allow the parties to "avoid . . . misunderstanding, delay, expense, and complexity," that can occur not just with litigation but with "prolonged administrative proceedings, or inaction."

The world of environmental conflict portrayed here is pragmatic and institutional, a world of delay, misunderstanding, inaction, potentially lumbering bureaucratic machinery, and political posturing. In this world parties with differing interests can nevertheless work together without compromising their "fundamental interests." This is not clearly a public world—one shared with past and future generations, one including the unorganized as well as those organized as parties to disputes at hand. Yet this is a world in which any mediation participant can pick up and go, so in this world parties somehow have viable alternatives to mediation—they are assumed to have enough power so that their participation in environmental mediation is not essentially involuntary.

Now consider the self-presentation of ICF, a self-described "Nationwide Professional Services Firm," in whose materials we find a crisp and informative portrayal of its "Regulatory Negotiation Services," a document worth quoting at some length:[4]

> Promulgating regulations is a fact of life for many government agencies. Increasingly however, regulations are challenged in court by outside interest groups, preventing or delaying their promulgation and implementation. To avoid such litigation, agencies are looking to regulatory negotiation as an effective means of gaining consensus on an approach to proposed regulations among divergent groups.

What Is Regulatory Negotiation?

ICF Incorporated—a nationwide professional and technical services firm—is able to help regulatory agencies develop rules that are well-accepted using a process called regulatory negotiation. This process involves bringing together representatives of key parties interested in a particular regulation to negotiate a rule that everyone can "live with." Whether the agreement reached between the parties is a general statement of principles or a detailed text of suggested regulatory language, the agency does not give up its regulatory responsibility or authority. What the agency gains, however, is substantial:

- New ideas for regulatory approaches the agency may not have considered;
- Agreements by traditionally opposed interest groups to cooperate and build consensus on key issues;
- Understanding of the types of regulatory approaches that are likely to be accepted;
- Awareness of areas of critical concern to various interest groups; and
- Exploration of each party's "bottom line," i.e. pushing the limit to which a party will negotiate before pursuing an alternative action, such as litigation.

What the other participants gain is equally substantial:

- An opportunity to focus on issues of greatest concern;
- Influence on decisions on how to balance and tradeoff among concerns;
- Better rules; and
- Avoidance of litigation.

Agencies that have used regulatory negotiation have found that the likelihood of a rule being challenged in court prior to promulgation is greatly reduced if the rule has had the benefit of such a negotiation process.

How Does Regulatory Negotiation Work?

The preliminary steps in initiating regulatory negotiation are critical and time consuming. In order for the consensus reached by participants in a regulatory negotiation to be effective, all interest groups must be adequately represented. Once parties are at the negotiating table, skilled facilitators must manage the negotiations toward consensus and ensure that any agreement reached is acceptable to all parties, including the agency that will promulgate the rule. Equally important is the facilitator's role in ensuring that representatives who sit at the negotiating table are communicating regularly and effectively with their broader constituencies. ICF has the capability to:

- Assist public agencies in planning and structuring the regulatory negotiation process;
- Identify and convene appropriate participants;
- Facilitate communications throughout the negotiation process, including those that take place between formally organized meetings;

• Provide all logistical and facilitation support necessary to prepare for multi-party, multi-session negotiations; and
• Provide the technical, scientific, and policy analysis expertise necessary to support the demands of a negotiated rulemaking.

These passages provide a prototypical example of the practical, public rhetoric—the self-presentation—of those providing regulatory negotiation services. (The truth or falsity of these claims is not at issue here.) For our purposes here, it hardly matters if this portrayal of regulatory negotiation services is typical of 25 or 75 percent of other regulatory negotiation providers. What we find portrayed wonderfully here is a complex of issues suggesting the more general challenges of regulatory negotiations. Notice, for example:

First, the firm promises that regulatory negotiation complements rather than replaces authoritative decision making: "The agency does not give up its regulatory responsibility or authority."

Second, the firm promises practicality and realism rather than idealism. We read no promises of "the best decisions" or the "right" decisions; instead we are searching for outcomes, for rules, that "everyone can 'live with.'"

Third, the firm promises that regulators and agencies will learn, gaining "new ideas for regulatory approaches the agency may not have considered" and "awareness of areas of critical concern to various interest groups."

Fourth, the firm promises access and influence to nonagency participants.

Fifth, the firm promises nonagency participants "better rules" (relative, presumably, to those promulgated if the agency were left to act on its own or take its promulgated rules to court to defend).

Sixth, the firm promises the avoidance of litigation, quickly noting that "the likelihood" of litigation is reduced when rules are negotiated.

The firm's characterization of the regulatory negotiation process is telling too. Strikingly, instead of the language of neutrality, we find the appeal to "skilled facilitators" and (toward the end of the document but not quoted above) "trained professionals who are experienced in convening and facilitating complex, multi-party, multi-issue negotiations."

Typically when claims to impartiality and neutrality are made, mediators and facilitators claim, as this firm does, to "assist public agencies in planning and structuring the . . . process." Yet here these professionals are portrayed as having responsibilities that reach far beyond mere "process facilitation," for they will "ensure that any agreement reached is acceptable to all parties" and "that representatives who sit at the negotiating table are communicating regularly and effectively with their broader constituencies." Not only that, they will also "identify and convene appropriate participants"!

To consider a third case, contrast ICF's presentation of the responsibilities of regulatory negotiation facilitation with the practical promises, the marketing rhetoric, of nonprofit CDR Associates' "Public Policy Program":[5]

While businesses, governmental agencies, and public interest groups often share mutual concerns, on occasion their objectives may be in conflict. Disagreements arise concerning public policy issues such as siting potentially controversial facilities (gravel mining operations, waste disposal plants, airports, and shopping centers), regulatory concerns, development plans and jurisdictional disputes. These disagreements can result in costly, time consuming and sometimes damaging conflicts that harm a community, state, or region.

CDR provides a forum in which groups with differing interests can plan and develop solutions to problems of mutual concern. Our professional staff organizes policy dialogues—forums where diverse interest groups can discuss and design mutually satisfactory policies. We also design the decision making process, provide experienced facilitation services, and offer logistical support to the participants. We have facilitated public policy dialogues on downtown economic development, county planning, transportation facility siting, social services policy, natural resources development, and governmental rule making.

These kinds of negotiations can involve many phases and require coordination among a variety of groups and participants. Intervention by CDR's process design specialists can result in a reduced timetable for agreement, more professional discussions, and better planning of events. The CDR process fosters a spirit of cooperation in formulating public policy issues, which so often take on a hostile, adversarial character.

The decision process begins by properly identifying all concerned groups and their objectives. A forum is established for dialogue so that each participant can develop an understanding of the concerns of the others. By jointly designing new and acceptable alternatives which address multiple interests, a decision can be reached that all parties support. The process generally requires less time; results in more amicable, cooperative working relationships; and produces economically, politically, and environmentally sound results.

Why CDR?

CDR intervention specialists have worked with diverse groups—from large industrial and governmental agencies to public interest groups—in resolving difficult public policy problems. The CDR process has proved to be effective. Each settlement program is tailored to the individuals and issues involved. Definition of the problem and identification of the parties are just the first steps in developing settlement options and arriving at solutions acceptable to all concerned.

If you or your organization are involved in a complex, multi-party issue, contact CDR for expert assistance in the design and implementation of effective decision-making procedures which can produce long lasting, mutually satisfactory results."

This text too should be read as a cultural document. The text conveys the style and language with which CDR publicly defines itself and portrays its services. CDR's staff are "process design specialists," with no apparent substantive commitments. The CDR process "has proven to be effective"; whatever "effective" means more generally, here its use means, "You can trust us." In the "policy dialogue" forums, "groups with differing interests" learn; they "develop an understanding of the concerns of the others." They "develop solutions to problems of mutual concern." Through the policy dialogue process, parties develop "a spirit of cooperation" and "more amicable, cooperative working relationships," and they reach "long lasting, mutually satisfactory results."

Claims to neutrality have been preempted here by presumptively authoritative claims to "process specialization." The reference to the challenge of "properly identifying all concerned groups" asks us to assume, apparently, that "process specialization" will do what needs to be done. We are in responsible hands.

Recurrent Themes

Each of these three public dispute-resolving bodies has shaped its claims to be attractive. These documents convey not simple statements of "the record" but rather self-defining, indeed self-constituting, expressions of competence, expertise, and responsibility. Their claims are not just marketing appeals designed to attract consumers, though they are that too, for these materials also work to persuade readers that public dispute-resolution processes have a certain character that these professionals will enact—a character of being practically effective, ethically virtuous, and politically responsible (Frug 1988).

These processes are practical, readers should understand, because the mediating and facilitating staff are professionals; the staff are trained and experienced specialists with proved track records. These professionals will supply "logistical support" to the parties; everyone's concerns will receive attention. Negotiated outcomes, too, it seems, may be more stable than those imposed by any one party.

These processes are virtuous, readers should learn, for they foster a cooperative rather than a hostile and adversarial spirit; they address mutual concerns; they free parties from distracting posturing; they generate new ideas; they improve relationships.

Finally, readers should be persuaded, that public dispute-resolution processes are politically responsible, for they are voluntary and not coercive; they complement, rather than preempt or threaten, established legal responsibility and authority. They are inclusive, seeking to give voice to all affected parties.

Now these claims to the good are just that: too good to be taken at face value. In the three documents, these claims work as metaphors rather than arguments, but they are powerful political metaphors nevertheless (Gusfield 1981, White 1985). In each case, our understanding of public, and not strictly private, matters is at stake—in particular, our understanding of legitimate public modes of conflict resolution and public participation.

These documents also argue that we can overcome the inadequacies of established planning, regulatory, and legal processes whose biases, inefficiencies, and inequities are all too familiar. Susskind and Cruikshank (1987:9), for example, crisply reflect the narrow scope of litigation in many cases where multiple interests are at stake: "Simply put, the court's purpose is to interpret the law, not to reconcile conflicting interests." So our understanding of deliberate intervention is at stake too—whether that intervention is based in a nonprofit firm, a public interest organization, a "professional services firm," or a public bureaucracy. The examples we have considered so far illustrate the problematic claims of public dispute-resolution providers, but my argument is not that public dispute resolution should be privatized. As we shall see, these claims reflect far more general challenges faced by public officials and community leaders alike.

Can mediator-facilitators, so-called intervention specialists, help us deal with public disputes in better ways than we conventionally might? If we are concerned with matters of public policy, democratic politics, and a safe and healthy environment, what kinds of public dispute-resolution practices should we encourage (Ury, Brett, and Goldberg 1988; cf. Laue and Cormick 1978)? What sorts of practices should we guard against (Blackburn 1988, Jaffe 1984, Luban 1985)? And why—for what reasons? To address these questions, this chapter explores the challenges of dispute-resolving practices of planners, public administrators, and managers in a wide variety of public agencies—city planning departments, federal, regional, and state environmental agencies, local departments of health and community development, housing authorities and school systems, and others.

Public Dispute Resolution, Planning, and Public Management

The mushrooming literature on public dispute-resolution practices provides all too little in the way of careful answers to these questions. The province of practitioners and political philosophers alike, these questions abide. Their answers, we can be sure, will have deeply practical implications.

Few authors, and far fewer practitioners, have written as cogently about the issues here as Lawrence Susskind. Susskind's work is striking because it helps us to pose, and begin to answer, crucial practical and political questions about the promise of mediation understood as an integral part of planning and public management practice. Susskind has most persistently addressed the special character of public sector dispute resolution (relative, for example, to conventional labor mediation) as well as called attention to the perplexing problems to be found there. His work is distinctive, furthermore, for its rejection of mediator neutrality and its call for "activist mediation." This chapter can be read as both an appropriation of Susskind's work (e.g., its concerns with the special responsibilities of public sector mediators) and a critique and perhaps refinement of it, particularly its heavily interest-based language (cf. Susskind and Field 1996).

Susskind and Ozawa (1983:257) point directly to distinctive and crucial features of public disputes: "Public sector disputes are special. They differ from conventional two-party private disputes in that they involve choices with substantial spillover effects or externalities that often fall most directly on diffuse, inarticulate, and hard-to-represent groups (such as future generations). It is our contention that mediators involved in resource allocation decisions in the public sector have responsibilities that transcend those facing mediators in more traditional situations."

Accordingly, Susskind's characterization of an "activist model" of mediation, for example, squarely challenges other proponents of public dispute resolution to move beyond thin descriptions of mediators as purely "process people." Noting that city planners face ubiquitous conflict, Susskind has argued that these professionals can play quasi-mediating roles and so foster fair, efficient, well-informed, and stable resolutions to public disputes. Disputes over rights, however, are *not* to be mediated (Susskind and Cruikshank 1987).

Such activist mediators, Susskind argues, are nonpartisan, yet they have several responsibilities: they must be concerned with both the representation of affected parties in the mediation *process* and the efficiency, stability, and well-informed character of potential mediation *outcomes* (Susskind and Cruikshank 1987, Susskind and Ozawa 1983, Forester and Stitzel 1989a). Susskind and Madigan (1984:202) articulate the activist mediation position boldly:

> In the final analysis, it is "active" mediation that is most likely to prove the key to success. Despite the debate surrounding active versus passive mediation, we believe active mediation is to some degree critical in almost any public sector negotiation. Such activism is perhaps most important when (1) the parties are not able to meet face-to-face without assistance; (2) several diverse interests should be represented at the table; and (3) participants lack the skills or knowledge necessary for communication, brainstorming, or joint problem-solving. In the end, the active mediator ensures that all parties have a chance to participate. The goal of the mediator should be to help the parties reach an agreement which they and the rest of the community will, in retrospect, view as wise and fair.

In sum, mediated negotiation may not be appropriate for all public decisions or disputes. But decision-makers and dispute resolvers should seriously consider mediated negotiation as a supplement to traditional administrative, legislative, or judicial processes, especially when the actual or potential conflict exhibits those characteristics described above.

Public sector mediators cannot be content simply to let organized parties meet and "make a deal," excluding weaker and perhaps unorganized parties. But the problems here are vexing (Susskind and Cruikshank 1987, Cormick 1987, Laue and Cormick 1978, Susskind and Ozawa 1983).

Identifying "representation" as one of the "obstacles to more widespread use of mediated negotiation in the public sector," Susskind and Ozawa (1983:274) stress that "one of the first hurdles to overcome . . . is the identification of all the parties likely to hold an interest in the outcome. In private disputes, the affected parties identify themselves. [But] in public disputes, especially those with spillover effects, the definition of legitimate stakeholding interests can itself lead to conflict." So the issue of representation or legitimate participation is posed here but hardly "solved."

In addition, to promote stable agreements, public sector mediators must work to anticipate "the steps necessary to ensure implementation." They must be concerned, too, with provisions for monitoring as well as for possibilities of renegotiation (Susskind and Ozawa 1983:268, Susskind and Cruikshank 1987). To promote well-informed or wise agreements, these mediators must be concerned that negotiations take advantage of the best available information—that the search for options be well informed, perhaps through the use of joint fact-finding efforts. To promote efficient outcomes, finally, the mediator's responsibility is to help the parties to a dispute to "exploit their differences"—to trade when they can—to achieve "joint gains," and thus to minimize any unrealized opportunities remaining to "give in order to get" (Susskind and Ozawa 1984, Raiffa 1985).

These recommendations are all practically pitched, and they deserve to be explored far more systematically than they have been (Burton 1988). If we wish to improve public dispute-resolution practice, we must explore further questions too: How can and should mediators of public disputes deal with these issues of representation? How should they help affected parties to listen, to reformulate issues, to learn? What character of interaction between the parties should these mediators encourage?

What commitments, as well as responsibilities, ought the facilitators or mediators of public disputes to have (Blackburn 1988)?

Yet before we can explore these issues, we must ask a prior question: When affected parties "participate," just what are they doing? If, in part, they are coming to see both issues and the perspectives of others in new ways, and coming to see their own priorities, "interests," and relationships—*themselves* and others—in new ways, then they are doing far more than "exploiting their differences," engaging in the trades of pluralist, interest-based bargaining. As we shall see, though, the idea of "exploiting differences" to achieve joint gains is a powerful idea (perhaps a *too* powerful idea), for it threatens to reduce the deliberative dimensions of political participation to its interest-based terms (Forester 1999b).

Mediators and facilitators typically do attend to formal procedures and ground rules to ensure that all parties have not just time to speak, but access to the process and safety from continued interruption or personal attack. But if these mediators also probe issues in ways that enable the parties to reformulate their own interests, needs, and well-being, then these euphemistically called "intervention specialists" are catalysts of public learning and discovery. These mediators and facilitators are public stewards, not just apolitical neutrals. They are organizers of public debate and deliberation, not just convenors who serve water and ask everyone to be polite. We need to ask then, as we shall, about the character of that public deliberation.

The practice of public dispute resolution appears to be light-years ahead of any "theories" purporting to characterize public dispute-resolution practice. That suggests both good and bad news. The bad news is that accounts by practitioners can be expected to be ambiguous, nonsystematic, long on symbolism and short on analysis, and hardly adequate to the political complexity and richness of their own practice.[6]

The good news is that relevant debates about public management and public deliberation, about pluralist interest group politics and conceptions of public good, about individual autonomy and social identity (in political philosophy), show us the political significance of public dispute-resolution efforts. We can explore these issues as they arise in the practice of city planners and public managers more generally.

Challenges of City Planning Practice

Several recent accounts suggest that public sector planners routinely work in the face of conflict and face mediating responsibilities as a result. Reviewing the limits of traditional formulations of planners' roles as distanced technicians or interest group advocates, Susskind and Ozawa (1984:9) argue, for example:

> In our view, implementation failures are a consequence of the planning profession's hesitancy to stress the important role that planners can play during implementation, especially in building and maintaining a durable consensus and in resolving disagreements. These tasks are central to our view of the planner as mediator. . . . The mediator planner does more than bring contending parties together around a common plan. The mediator-planner commits to a continuing role, seeking to ensure that the interests of all parties affected by a policy or a plan are taken into account from beginning to end.

Why might the idea of planners acting as mediators even be plausible? Consider the situations in which many local planners find themselves. These planners often have complex and contradictory duties. They may seek to serve political officials, legal mandates, professional visions, and the concrete requests of particular citizens' groups all at the same time. They typically work in situations of considerable uncertainty, great imbalances of power, and multiple, ambiguous, and conflicting political goals. Such local planners may seek ways to negotiate effectively, as they try to satisfy particular interests, *and to mediate politically too,* as they try to resolve conflicts through a semblance of a participatory planning process (Forester 1987a, 1989).

But these two tasks—negotiating and mediating—appear to conflict in two ways. First, the interestedness of a negotiating role threatens the independence and presumed neutrality of a mediating role. Second, although a negotiating role might allow planners to protect less powerful interests, a mediating role threatens to undercut this possibility and thus to leave existing inequalities of power all too intact (Paris and Reynolds 1983).

How, then, can local planners and public managers who seek not only to be responsive to affected citizens but to be effective, too, employ whatever discretion they have to deal with these problems (Healey et al. 1988, Bryson and Crosby 1992)?

Challenges of Public Management Practice

In a striking essay, Robert Reich argues that students of politics and policymaking often misunderstand their own administrative discretion. "The postwar transformation of public administration," he tells us, "centered on two related but conceptually distinct procedural visions of how public managers should decide what to do. The first entailed *intermediating among interest groups;* the second, *maximizing net benefits*" (Reich 1988: 129). Behind the first lies pluralist political theory; behind the second lies decision theory and microeconomics.

Reich shows how poorly these views explain the actual work of public managers, planners, public administrators, and, as we shall see, public dispute resolvers. We quote Reich (1988:138) at some length, noting that public managers and planners typically confront many of the challenges faced by environmental mediators and other professionals working to resolve complex public disputes:

> Both interest group intermediation and net benefit maximization share a view of democracy in which relevant communications all flow in one direction: from individuals' preferences to public officials, whose job it is to accommodate or aggregate them. . . . This view of the place of public management in a democracy suffers from two related difficulties. First, . . . individual preferences do not arise outside and apart from their social context, but are influenced by both the process and the substance of policy making. The acts of seeking to discover what people want and then responding to such findings inevitably shape people's subsequent desires. . . . [Second, this view] leaves out some of the most important aspects of democratic governance, which involve public deliberation over public issues and the ensuing discovery of public ideas.

The intermediator, Reich argues, shapes the public framing of problems at hand in part by shaping the participation of affected parties. Managing that participation can affect the reputation and public standing of those parties. They may now be seen as having access to power, be seen publicly as personages to be taken more seriously. Inclusion may dignify some actors but not others. The public manager shapes public attention selectively. These issues, now demonstrably on the public agenda, may be taken more seriously; those other issues may be relatively neglected. As organizers of attention, public managers mobilize bias, as do all other organizations, as Schattschneider taught long ago (Schattschneider 1960, Majone 1989).

"In all these ways," Reich (1988:140–141) sums up, the public manager as "interest group intermediator is an active participant in the political development of the community. By recognizing "established" groups and leaders, and subtly encouraging others to participate, the intermediator effectively shapes public understandings of what is at stake, perceptions of who has power in the community, and assumptions about what subjects merit public concern. In this way, he alters the political future. To view him merely as a neutral intermediator dramatically understates his true role."

Reich's analysis speaks directly to the practical challenges faced by mediators or facilitators of public disputes. Like public managers, they are not neutrals; they too face issues of assessing and managing the adequate representation of affected parties. Like public managers, they too must be sensitive to the ways issues are defined and redefined, posed and reformulated (Pops and Stephenson 1988). Most significant, they face the challenge that Reich finds central to public management properly understood: they too must nurture a process of public deliberation and learning, a process of civic discovery (Kemmis 1990, Mathews 1994).

Reich (1988) develops the point roughly as follows:

> Imagine now that the public manager eschews both types of conventional policy making. Instead of assuming that he must decide [upon a recommended course of action], he sees the occasion as an opportunity for the public to deliberate over what it wants. . . . He then encourages . . . the convening of various forums—in community centers, schools, churches, and workplaces—where citizens are to discuss whether there is a problem and, if so, what it is and what should be done about it. The public manager does not specifically define the problem or set an objective at the start. . . . Nor does he take formal control of the discussions or determine who should speak for whom. At this stage he views his job as generating debate, even controversy. . . . In short, he wants the community to use this as an occasion to debate its future. . . .
>
> Several different kinds of civic discovery may ensue. . . . The problem and its solutions may be redefined. . . . Voluntary action may be generated. . . . Preferences may be legitimized. . . . Individual preferences may be influenced by considerations of what is good for society. . . . Deeper conflicts may be discovered. . . . Deliberation does not automatically generate these public ideas, of course; it simply allows them to arise. Policy making based on interest group intermediation or net benefit maximization, by contrast, offers no such opportunity. (pp. 144–146)

There is a deep resonance here between the deliberations to be fostered by the public manager and the qualities of interaction and participation centrally at issue in dispute-resolution practices. A purely interest-based negotiation process would be subject to the same criticisms that Reich and other political theorists bring to bear against interest group intermediation. If mediators and facilitators treat parties' preferences and interests as if they were pregiven, they will fail to promote a process that helps parties to learn, to refine their initial preferences, to see anew issues, options, and perhaps their actual interests as well. To promote public deliberation that encompasses but extends beyond "interests," mediators and facilitators need a richer account of public disputes and public participation than an interest-based pluralist political theory offers.

Indeed, recognizing that we refashion ends and means together in processes of political interaction, dialogue, and professional practice too, social and political theorists have recently explored the character of deliberation and practical judgment (Rorty 1982, Bernstein 1983, Schön 1983, Wiggins 1978, Kemmis and Carr 1986, Nussbaum 1990b). We must extend that work if we ever wish to ask, "What sort of public deliberation should public policy promote? What sorts of public deliberation should planners and policy advisers encourage?"

In the next section, we begin to offer such an alternative account—a modern republican conception of political conflict, political participation, and political life—that can help us to understand and improve public dispute-resolution practice. But before turning to that account, we consider a final irony, the potentially weak legitimating force of pluralist public management. Reich (1988:146–147) warns,

> The failure of conventional techniques of policy making to permit civic discovery may suggest that there are no shared values to be discovered in the first place. And this message—that the "public interest" is no more than an accommodation or aggregation of individual interests—may have a corrosive effect on civic life. It may invalidate whatever potential exists for the creation of shared commitments and in so doing may stunt the discovery of public ideas [in turn calling into] question the inherent legitimacy of the policy decisions that result. For such policies are then supported only by debatable facts, inferences, and trade-offs. They lack any authentic governmental character beyond accommodation or aggregation.

Recall, too, Lindblom's (1959) wonderful conundrum discovered in his defense of incrementalism: if it is simply agreement that is the measure of a decision's goodness, then decision makers are put in the position of defending any agreed-on decision as good even if they cannot say what the decision is good for!

This argument suggests that public dispute resolution will be vulnerable to being seen as deal making unless its advocates and practitioners can articulate a broader conception of political life, the public interest, and the textures of the public goods of deliberation and debate, mutual recognition and discussion, learning and civic discovery. Yet these political values are typically ignored in strictly interest-based accounts, for they characterize not the issues in dispute but the encompassing and even constituting political process of dispute resolution itself. These values can be understood as second-order interests: meta-preferences in Hirschman's (1986) terms, self-commands in Schelling's (1984) terms, or precommitments in Elster's (1979) terms. No doubt invoked by mediators to "stay on track," "to remind everyone why we're here," these values or second-order interests are unlikely to be articulated by many disputants in any one case, even if commitments to these values are arguably at stake in all public interaction (Alexander 1989).

The problem with the pluralist rejection of "the public interest" is not that a "unitary public interest" can be "discovered," but that the problem or question is poorly posed in the first place: a collective political problem of public-ness is posed in an intentionalist language of interested-ness. Seeking to find the true measure of "the" public interest, accordingly, is a bit like yearning to learn the intimate secrets of immaculate conception. It is really not something we can, or ought to try, to do.

We need instead to characterize the public realm in which we wish to live and act together. Posed this way, the "problem of the public interest" becomes more tractable. No longer faced with discovering what cannot exist as an intentional interest in the first place, we now find it obvious that *no* one quality (Safety! Equality! or Community!) will characterize sufficiently what we must aspire to if we hope to preserve a public realm of political interaction in which plurality, difference, and diverse conceptions of "the good life" can endure. Freeing ourselves from the economistic language of public interest, we can turn to the inescapable task (shared by all pluralists!) of fostering a safe and enabling public realm.

We need, as chapter 5 suggested, to investigate the participatory rituals of public dispute-resolution processes as they shape social and political interactions. In such rituals of storytelling and small meetings, we and others reproduce or transform our identities. Witnessing an interruption in a meeting, for example, we may think, "Well, he's not so passive after all!" (identity transformed) in one case, or in another, "She *really* is obnoxious!" (identity confirmed). Although participatory rituals can be integrative or transformative, we know all too little about how they may be so in public dispute-resolution processes. We should do research to find out if some rituals (greetings, turn taking?) may not be relatively integrative, while others (role taking?) might have more transformative possibilities in public dispute-resolution contexts. We should expect, too, that the efficacy and significance of particular rituals will vary, just as disputants' "cultures" will vary (Kertzer 1989).

Reich's reflections serve us in several ways. First, he suggests that a pluralist political account inadequately captures the developing interplay of political interests and identities at stake in public disputes (Tribe 1973: 634, 656). Second, he suggests that a pluralist account begs questions about the orchestration of representative participation: unless interest groups are pregiven, preorganized, unless every affected party has its Lindblomian "watchdog," then the public manager and the public dispute resolver cannot merely be neutral intermediators between given parties (Lindblom 1959, 1990, Forester and Stitzel 1989a, Smith 1985). Third, he suggests that pluralist interest-based bargaining by itself will raise questions of legitimacy, for deals struck may maximize joint gains efficiently and yet be perceived publicly as illegitimate (Burton 1988:135, Susskind and Cruikshank 1987, Forester 1985a).

Public Deliberation and Participation: Republican Political Theory Appropriated

Once the environmental mediator, regulatory negotiation convenor, or policy dialogue facilitator makes judgments about and manages "appropriate participation," that mediator-facilitator has assumed a public-creating responsibility—a constituting or constitutional responsibility to structure public interaction, debate, and processes of dispute management or settlement. In this light, the "process design specialists" who structure policy dialogues for CDR Associates are aptly, if technocratically, named. But, some will ask, who are these "process design specialists" to constitute the nature of public conflict—as, let us say, a matter of interests, or one of rights, or perhaps as one of traditional values? Silbey and Sarat (1989) show clearly that our understanding of public disputes (as claims? grievances? expressions of need?) is a matter of political struggle. In practice mediator-facilitators create political spaces in which conflicting parties speak and listen, recollect their experiences and express their needs, articulate their interests, and invoke their commitments.

This constitutive influence of mediators enacting "public dispute-resolution processes" raises substantial research questions about institutional design (Elkin 1987:105–109). Carpenter and Kennedy (1988:29) write, for example, "The goal of consensus decision making is to reach a decision that all parties can accept. . . . Not everyone will like the solution equally well . . . but the group recognizes that it has reached the best decision for all parties involved. *They do not vote, because decisions made by voting create winners and losers,* and disgruntled losers can cause problems later on by challenging the decision" (emphasis added). Just how does this "creation" of winners and losers work in voting processes? How in mediated processes are alternative identities created and shaped? What identities should be shaped in public dispute-resolution processes: identities as private persons, territorial residents, interest maximizers, communitarian loyalists, public citizens? We have too little research that informs these questions—notwithstanding claims by alternative dispute-resolution proponents to be creating collaborative relationships and, presumably, parties.

Fulfilling this public-creating responsibility requires mediator-facilitators to *protect* and *nurture* an always precarious and contingent democratic public sphere—not only a marketplace for exchange but a public space in which citizens both meet together and continually refashion themselves. But how is this possible?

Frank Michelman (1988:1500) argues that American constitutionalism rests on two fundamental premises: that the American people are, first, self-governing and yet, second, governed by laws and not people. How, he asks, can we have both the government of the people by the people and the government of the people by laws? Precisely the same tension exists in most public dispute-resolution processes: the processes are rule defined, with prerogatives of parties to be protected, yet they are also managed, necessarily nurtured, by a mediator or facilitator. Michelman (1988:1499) notes Margaret Jane Radin's Wittgensteinian argument: "Radin concludes that 'if law cannot be formal rules, its people cannot be mere functionaries.' The flip side of that is that if the people are not mere functionaries (but rather are self-governing), then their law cannot be formal rules." So too for the character of the "rules" governing public dispute-resolution processes.

This public sphere has been known classically as *res publica,* the object of both classical and modern republican political theory. Yet meeting, arguing, venting, listening, and often learning in this sphere, participants fashion outcomes and refashion themselves, their reputations, their senses of future possibility, and their relations with others as well (B. Smith 1985:18).

Accordingly, the discussion that follows reflects recent attempts to reappropriate republican political thinking to retain its sensitivity to notions of public life, civic discovery, citizenship, and self-transformative political participation, while freeing such understandings of their earlier linkages to exclusionary and repressive practices. Sunstein (1988:1539–1540) writes, for example,

> Various strategies of exclusion—of the nonpropertied, blacks, and women—were built into the [classical] republican tradition. The republican belief in deliberation about the common good was closely tied to these practices of exclusion; it cannot be neatly separated from them. In some of its manifestations, moreover, the republican tradition has been highly militaristic, indeed heroic—exalting, as the model for public life, the fraternity of soldiers during wartime. Equally important, the republican belief in the subordination of private interests to the public good carries a risk of tyranny and even mysticism. The belief is also threatening to those who reject the existence of a unitary public good, and who emphasize that conceptions of the good are plural, and dependent on perspective and power.

If we seek to understand and improve public dispute resolution, we must explore three closely related political questions. First, we need to understand the public world at stake in the process of public dispute resolution. Second, we need to characterize the "disputants," "affected parties," and "participants" in the process. Are they citizens, for example, or consumers, bundles of private interests (Sullivan 1986:157–159)? Third, we need to understand the possibilities of processes of political deliberation and their vulnerability to insinuations of power. Only then can we turn to the political and ethical question: What should the mediator-facilitators of public dispute-resolution processes seek to do? What sort of person should they try to be? We consider each of these questions in turn.

The Reconstructable Public World of Public Dispute Resolution

Concerned with the interest-based pluralist neglect of our public world, William Sullivan (1986) writes,

> As a liberal [utilitarian] would see it, the institutions of our society, the government, corporations, even the cities in which people live and work are all variations on the model of a business enterprise. Political, even social, vitality and progress are measured according to economic criteria. The public good, seen that way, becomes the utilitarian sum of individual satisfactions. A common interest can be presumed to lie only in ensuring advantageous conditions of general exchange—what is called, more realistically, a "good business climate."

Sullivan challenges advocates of interest-based negotiations to envision public participation that offers not just a climate for good business, but membership in a community of mutual concern, citizenship in a political community with others. In public dispute-resolution processes, then, we need to build not simply marketplaces for exchange, but democratic public spheres—settings where citizens can speak and listen, argue and negotiate, come into conflict and yet act together too. At stake here is recognizing what Michelman (1988:1496) calls the self-transformative capacity of our political communities. Failing to study and understand how public dispute processes may reconstitute political spaces, issues, and identities alike, we will remain blind to both possibilities we can realize and dangers we should certainly avoid. Too often, for example, as Carpenter and Kennedy (1988:16) note, "complex public disputes can become sinks for resources that the parties never meant to commit." Too frequently, too, as Harter (1989:2) writes, when participation is restricted to shaping "the record" and the facts of the matter, "the facts become a surrogate for a political process, and they are often corrupted as a result."

Drawing lessons from the civil rights movement, Michelman (1988: 1531) argues,

Much of the country's normatively consequential dialogue occurs outside the major, formal channels of electoral and legislative politics, and . . . in modern society those formal channels cannot possibly provide for most citizens much direct experience of self-revisionary, dialogic engagement. Much, perhaps most, of that experience must occur in . . . what we know as public life, some nominally political and some not: in the encounters and conflicts, interactions and debates that arise in and around town meetings and local government agencies; civic and voluntary associations; social and recreational clubs; schools public and private; managements, directorates, and leadership groups of organizations of all kinds; workplaces and shop floors; public events and street life; and so on. Those are all arenas of potentially transformative dialogue.

Mediated negotiations and public deliberations, policy dialogues and negotiated rule-making processes can also be such potentially transformative dialogues—dialogues that have a public yet not necessarily state-centered place. Here, too, the rules of the public dispute-resolution game will matter, for if weak but affected parties are not represented in negotiations, the ensuing process becomes not self-transformative but self-aggrandizing, reflecting not the inclusiveness of a public realm but the exclusivity of private power. Just as republican views of self-governance must embrace conceptions of the rule of law, as we shall see, so must our views of mediation and related deliberative practices address the ethics of mediator accountability (Susskind 1981, Stulberg 1981, Susskind and Ozawa 1983, Forester 1999b, Lowry, Adler, and Milner 1997).

In planning and public policy settings, what counts is not simply agreement, but the quality of any negotiated agreement. Susskind and Ozawa (1983:277) write, significantly,

Our objective in this article has not been to advocate the use of mediated negotiation in public disputes, but rather to urge its proponents to consider seriously whether mediators can be held sufficiently accountable to the interests of the public at large. In our view, mediators might be sufficiently accountable, but only if (1) they choose an appropriately activist model to guide their practice (. . . definitely not the labor mediation model); (2) they adopt an appropriate credo . . . known to all potential participants . . . ; (3) they assume measures of success that emphasize the quality (but not the particular substance) of agreements; and (4) they continue to seek better ways of overcoming the obstacles to more widespread use of mediated negotiation in the public sector.

Participation in the Modern *Res Publica:* Beyond Interest-Based Bargaining

When "affected parties" participate in public dispute-resolution processes, what is at stake? Exploiting their differences to be sure, but anything more? Far more. We must both take opportunities to "achieve joint gains" and recognize and seize further opportunities as well. Hanna Pitkin (1981:347) helps us here, for she shows powerfully how a modern republican understanding of public participation illuminates issues reaching far beyond the province of private interest satisfaction:

> We come to politics with our private interest firmly in hand, seeking by any means necessary to get as much as we can out of the system. . . . But actual participation in political action, deliberation, and conflict may make us aware of our more remote and indirect connections with others, the long-range and large-scale significance of what we want and are doing. Drawn into public life by personal need, fear, ambition or interest, we are there forced to acknowledge our power and standards. We are forced to find or create a common language of purposes and aspirations, not merely to clothe our private outlook in public disguise, but to become aware ourselves of its public meaning. We are forced, as Joseph Tussman has put it, to transform, "I want" into "I am entitled to," *a claim that becomes negotiable by public standards.* In the process, we learn to think about the standards themselves, about our stake in the existence of standards, of justice, of our community; so that afterwards we are changed. Economic man becomes a citizen. (emphasis added)

Pitkin's account is instructive both normatively and descriptively. Normatively, she characterizes the public learning at stake in processes of political participation—when, for example, citizens participate in facility siting decisions, policy dialogues, or regulatory negotiations. Descriptively, she points to research questions we should pursue in the study of those same public dispute-resolution processes. Shouldn't we investigate, for example, the transformations of awareness and recognition, of self and other, of private and public, to which Pitkin calls our attention (Abers 1998, Sunstein 1988, Rosen 1989)? These reflections suggest a deep point: political deliberation involves far more than marketplace trading and interest-based bargaining, for in such deliberation, minds change; we learn; we come to see the interests and concerns of others and ourselves anew; we come to recognize and perhaps appreciate previously unrecognized values; new patterns of significance arise; and so on.

Beth Meer (1989) presents the most sustained and suggestive analysis of these issues as they arise in environmental mediation, for she shows the complementarity of normative deliberation and pluralist interest-based bargaining, and exposes the blinders of the latter view of environmental mediation. Fully sensitive to issues of value conflict, Meer argues nevertheless that in normative deliberation (part of environmental mediation, properly understood) parties may not only shift their focus from positions to interests, but from positions to values as well and in so doing not only reconcile norms and interests but discover new values too. In Fisher and Ury's (1982) compelling phrasing, interest-based bargaining is a meta-"position" of its own, distracting our attention from the opportunities we have to satisfy other meta-"interests" (moral and aesthetic concerns)! We need further research to explore these issues—the nature of the continuum linking "fundamental values" to "interests" or "wants," and corresponding implications for negotiation or reconciliation, for example—as they arise in public dispute-resolution practice (Forester 1999b).

If an interest-based model of public dispute resolution is too limited, a less economistic and more political, modern republican model might both do justice to expressions of private interests and help appreciate and promote public deliberation. Cass Sunstein (1988:1548) poses the problem this way:

> The function of politics [in republican thought] is not simply to implement existing private preferences . . . , to aggregate private preferences or to achieve an equilibrium among contending social forces. . . . The republican position is instead that existing desires should be revisable in light of collective discussion and debate, bringing to bear alternative perspectives and additional information. Thus, for example, republicans will attempt to design political institutions that promote discussion and debate among the citizenry; they will be hostile to systems that promote law-making [and public agreements] as "deals" or bargains among self-interested private groups.

But now, if parties to public dispute-resolution processes not only construct agreements but reconstruct themselves—in part as a result of being exposed to new information, in part as a result of the constellation of participants—then the *political significance and power* of the mediator-facilitator's role is more important to understand than ever before. Given their discretion, their need to make practical judgments throughout the dispute-resolution process, their hand in shaping representation and subsequent participation too, how should we understand what these perhaps "activist" mediator-facilitators are to do (Susskind and Madigan 1984: 202)? What conception of public interaction might provide the background for their practical judgments? How can we address the liberal pluralist's suspicion that republican political theories—daring to speak of participants coming precariously to see themselves anew—may slide quickly into tyrannies of moral majoritarianism (Sunstein 1988:1541, Gutmann 1985, Taylor 1989a)?

We are caught between the proverbial rock and a hard place here. Encouraging participation in public dispute-resolution processes, we should hope and work for far more, as Pitkin teaches, than marketplace trading. We can seek to "exploit differences," but we can do even better, enriching parties' understandings of others, themselves, and their common public world and its possibilities. Yet we certainly do not want to encourage public participation that coerces parties to develop some one party's privileged "understandings of others, themselves, and their common public world and its possibilities." So what are we to do?

Fostering Political Deliberation and Resisting the Insinuations of Power

By attending to the precarious possibilities of political deliberation, modern republicans suggest, we can inform both our vision and our practice of public dispute resolution. Sunstein (1998:1549–1551) argues,

> The republican belief in deliberation is aspirational and critical rather than celebratory and descriptive. It is a basis for evaluating political practices. Modern republicans do not claim that existing systems actually embody republican deliberation. The republican commitments may reveal that actual deliberation, and purportedly deliberative processes, are badly distorted. . . . Often said to be antagonistic to private rights, republican theories are not, however, hostile to the protection of individual or group autonomy from state control. . . . What is distinctive about the republican view is that it understands most rights as either the preconditions for or the outcome of an undistorted deliberative process.

Ury, Brett, and Goldberg (1988) too suggest that conflicts of interests lie within a larger domain of rights, whose actualization and enforcement depend in turn on encompassing relations of power. What this formulation gains in elegance, it loses in conceptual clarity and insight, because the expression, articulation, and even constitution of interests not only depend on, but reveal in deed, the relations of power that parties bring to the negotiating table.

Processes of deliberation are historically contingent: subject, for example, to the problems of bias and agenda setting, to inequalities of knowledge and resources, and perhaps also to the cultural or linguistic privileging of one mode of discourse (cool, logical, technical?) over another (hot, emphatic, symbolic?). Critics argue, "The civic republican aims of participatory democracy are . . . undermined by economic inequality. Economic need compromises the independence that civic republican theorists deem a prerequisite to civic virtue and produces political inequality" (Brest, 1988:1626–1627; cf. Kochman 1981). Participants in public dispute-resolution processes can never be guaranteed that those deliberative processes will be undistorted, so we must always imagine these processes lying within a backdrop of rights, and thus most likely of judicial or administrative review. Thus, while a virtue of the civic republican tradition is its appreciation of concrete deliberative processes, several authors take pains to argue that such rich deliberative processes must be set within a framework of rights (Sullivan 1986:160, Sunstein 1988, Waldron 1988).

Still, to see that deliberative processes come with no guarantees is a bit like seeing that we are mortal. We need to bemoan less the facts of our finitude and ask more how we can create rich deliberative processes in the time we have, in the settings we face.

Douglas Amy's (1987) sustained political critique of environmental mediation illustrates the practical point here. Concerned with the seduction of the inexperienced by the status and ploys of the corporate elite, and with the imbalances of resources and expertise that affected parties bring to many environmental disputes, Amy charges that supposedly collaborative and cooperative mediated negotiation processes have been quite biased instead—to the detriment of citizens who are relatively weak, perhaps not yet even organized, perhaps not even yet cognizant of their vulnerability to the distant decisions of others. Amy argues, for example, that mediation processes are vulnerable to the professional self-promotion of the mediators and that issues of collective significance may be fragmented as mediation processes treat individual cases. Although Amy does not spell out the practical implications of his critique, he does imply that legal redress is often a viable institutional option for the weak and unorganized. Curiously, though, legal institutions are also vulnerable to many of Amy's own criticisms: the legal profession can be self-promoting; the courts can fragment issues, and so on!

If this much is bad news, the rest of the news is worse, for certainly our traditional institutional processes of dispute resolution—legal redress or political organizing, for example—are *also always contingent,* vulnerable to the effects of resource inequalities and power imbalances. So whether it makes political sense for a citizens' group, for example, to join a mediated negotiation will always depend on the nature of their alternatives. Will the deliberation likely to be staged in court deal with the concerns at hand, or be shaped by the strength of an adversary's legal counsel?

Providing no recipes, Susskind and Ozawa (1984:14) want us instead to recognize the practical contingencies here: "In some instances, groups out of power or at the bottom of the income scale have achieved more through mediated negotiation than by direct action or litigation. Mediated negotiation can improve the position of disadvantaged groups in certain important ways. In contrast to litigation and other adversarial methods of dispute resolution, mediated negotiation emphasizes joint fact-finding and information sharing. Since a lack of information is often a major weakness of less powerful groups, especially in technologically sophisticated disputes, increased access to information may represent a significant gain." They go on to make a significant practical point: "The extent to which information is actually shared may depend on the ability of the mediator."

The challenge we face as we seek to promote full-bodied public participation and deliberation in public dispute resolution processes is to envision the goods of deliberation no more than to anticipate and work to counteract the insinuations of power that threaten the health and freedom of those processes. Accordingly, we must ask: How can we defend and nurture political participation in a way that protects plurality, in a way that does not succumb to the dangers of exclusion and the continuing dominance of those already relatively powerful?

When are we to say that the result of a policy dialogue is legitimate? Surely not when the only participants to the "dialogue" have been industry insiders with a token environmentalist invited along for the ride.

When should we say that an agreement produced in an environmental mediation is legitimate? Surely not when one party manages simply to impose its will on others, who go along simply because they had no other alternative (Coleman 1983).

These questions echo those of Frank Michelman (1988) who asks when a political process, given plurality, can validate a law as truly self-given. If any one faction of a community—industrialists, preservationists, environmentalists, "ists" of any other kind—simply imposes its will, we would have good reason to doubt the legitimacy of any resulting "agreement."

Michelman's argument is worth following closely. A plural political process—and here we include policy dialogues, negotiated rule-making processes, and broader public dispute-resolution processes—can legitimately produce law if three conditions (Michelman 1988:1526) hold. First, participation in the process must reflect "some shift or adjustment in relevant understandings on the parts of some (or all) participants." Call this the *self-transformative condition:* learning takes place. Second, there must exist "a set of prescriptive social and procedural conditions such that one's undergoing, under those conditions, such a dialogic modulation of one's understandings is not considered or experienced as coercive, or invasive, or otherwise a violation of one's identity or freedom." Call this the *noncoercive condition:* participants are empowered to act together. And third, the first and second conditions "actually prevailed in the process" (historically in the case of earlier agreements, or currently). Call this the *material condition:* the participation at stake is free from domination.

The second, noncoercive, condition is crucial here, for without it we could hardly distinguish—even in principle, whatever the practical difficulties—more from less coercively reached agreements. Michelman's discussion (1988:1528) of this basic condition of republican dialogue allows us to draw together several of the central ideas we have considered above:

> What [condition 2] apparently describes is a process of personal self-revision under social-dialogic stimulation. It contemplates, then, a self whose identity and freedom consist, in part, in its capacity for reflexively critical reconsideration of the ends and commitments that it already has and that make it who it is. Such

a self necessarily obtains its self-critical resources from, and tests its current understandings against, understandings from beyond its own pre-critical life and experience . . . communicatively, by reaching for the perspectives of other and different persons. . . . These dialogic conceptions of self and freedom are implications of the republican—the American—ideal of political freedom in a modern liberal state. Thus might a modern republican conception of political freedom make a virtue of plurality.

Michelman's is a vision of participants (in public dispute-resolution processes, for example) who bring together and reconstruct their understandings and expectations, their interests and commitments, their fears and suspicions, as they employ their capacities to listen and to learn, to confirm their hunches or come to see issues anew. In this view, mediation or policy dialogue sessions are certainly no longer marketplaces for trading across interests—so many dollars for so many square feet of land (developed or protected).

The point here is less to reject interest-based bargaining than to appropriate it as an aspect of a broader public process, a modern republican process, of dispute resolution. Thus, Fisher and Ury's compelling yet deceptively simple formulation of interest-based bargaining should be read as a heuristic guide, not as a systematic argument for an interest-based pluralist political philosophy of public dispute resolution. A republican conception of public dispute resolution can appropriate and move beyond the insights of interest reconciliation (e.g., exploiting differences to reach joint gains) without being limited to a thin pluralist politics. At least since Durkheim, sociologists have known that interest-based bargaining presupposes normative foundations and identities (Granovetter 1985). Recent work too assesses the importance of perceptions of fairness in economic exchange (e.g., Kahneman, Knetsch, and Thaler 1986).

In this view, public disputants do not only exchange; they argue and learn, search for new possibilities, create agreements, and in part remake themselves in the process (Burns 1989, Bowles and Gintis 1986, Kertzer 1989). In this view too, the conception of the public good is one of the good of plurality, the good of a respect for difference. Benhabib (1985: 348–349) amplifies the point:

By plurality I . . . mean . . . that our embodied identity and the narrative history that constitutes our selfhood give us each a perspective on the world, which can only be revealed in a community of interaction with others. . . . A common

> shared perspective is one that we create insofar as in acting with others we discover our difference and identity, our distinctiveness from, and unity with, others. The emergence of such unity-in-difference comes through a process of self-transformation and collective action. . . .
>
> Through such processes we learn to exercise political and moral judgment. We develop the ability to see the world as it appears from perspectives different from ours. Such judgment is not merely applying a given rule to a given content. In the first place it means learning to recognize a given content and identifying it properly. This can only be achieved insofar as we respect the dignity of the generalized other, who is our equal, by combining it with our awareness of his or her concrete otherness. What we call content and context in human affairs is constituted by the perspectives of those engaged in it.

These are the possibilities to which a modern republican account of political deliberation and participation alerts us. They provide the beginning of a research agenda for our continuing exploration of public dispute-resolution processes and practices.

Toward a Research Agenda I: Sensitivity to Power and Judgments about Exclusion

If mediators are to foster plurality, foster the virtues of a rich republican politics, and nurture transformative public debate, they must be able to anticipate and somehow respond to the play of power, including "those *social constraints* under which action [and deliberation] take place and self-identities are constituted" (Benhabib 1985:349–350). Listening closely, they must be sensitive to the problematic roots of parties' identities in community, ethnic, religious, and historical attachments (Marris 1975).

Inescapably, the mediators and facilitators of public disputes will have to wrestle with issues of inclusion and exclusion. They will have judgments to make about who is to count as an "affected" party, about who is to be represented, and who is to participate. General appeals to collaboration, joint problem solving, and dialogue simply beg the questions of inclusion and representation (Benhabib 1989:154, Mouffe 1993, Young 1990).

To help us formulate a research agenda to explore these issues, Seyla Benhabib's reflections (1989:155) about the promises and problems of public deliberation are worth quoting:

Social struggles both extend the participants of the public realm and extend the scope of the conversation in the public realm. It is therefore important that we cease thinking of the public realm solely in terms of . . . legislative or state activity. The original sense of res publica is the "public thing" that can be shared by all. Sharing by all means first and foremost that certain issues become matters of public conversation, and that, in Hannah Arendt's terms, they leave the sphere of private shame, embarrassment, silence, and humiliation to which they have been confined. In making public those issues and relations that had condemned us to shame, silence, and humiliation, we are restoring the public dignity of those who have suffered from neglect. . . .

[For] power is not only a social resource to be distributed, say, like bread or automobiles. It is also a sociocultural grid of interpretation and communication. Public dialogue is not external to but constitutive of power relations. . . . There are officially recognized vocabularies in which one can press claims; idioms for interpreting and communicating one's needs; established narrative conventions for constructing individual and collective identities; paradigms of argumentation accepted as authoritative in adjudicating conflicting claims; a repertory of available rhetorical devices (Fraser 1986, p. 425). These constitute the "meta-politics of dialogue," and as a critical theorist I am interested in identifying the present social relations, power structures, and sociocultural grids of communication and interpretation that limit the identity of parties to the public dialogue, that set the agenda for what is considered appropriate or inappropriate matters of public debate, and that sanctify the speech of some over the speech of others as being the language of the public. I believe the res publica can be truly identified only after the unreasonable constraints on public conversations have been removed.

Benhabib's remarks pose a research agenda for the analysis of power and structure, inclusion and exclusion in public dispute-resolution processes. What social or environmental "struggles" affect mediator-facilitators' thinking about representation or agenda setting? How to identify, include, or represent inarticulate groups—those who are perhaps silent before lawyers, humiliated before doctored experts, to say nothing of those too lacking in resources or hope to fight city hall? How vulnerable are mediators to being seduced by style, language, and familiarity in the way that even local planners may be?

These dangers are vividly reflected in a planning director's comments (quoted in Forester 1989:86) about his relationships to the parties involved in land use conflicts:

It's easy to sit down with developers, or their lawyers. They're a known quantity. They want to meet. There's a common language—say, of zoning—and they know it, along with the technical issues. And they speak with one voice (although that's not to say that we don't play off the architect and the developer at times—we'll push the developer, for example, and the architect is happy because he agrees with us . . .).

But then there's the community. With the neighbors, there's no consistency. One week one group comes in, and the next week it's another. It's hard if there's no consistent view. One group's worried about traffic; the other group's not worried about traffic but about shadows. There isn't one point of view there. They also don't know the process (though there are those cases where there are *too many* experts!). So at the staff level (as opposed to planning board meetings) we usually don't deal with both developers and neighbors simultaneously.

How should mediators handle differences of language, as these reflect not simply matters of translation but status, power, deference? Meer, for example (1989:chap. 8, citing Barber 1984; cf. Kochman 1981), discusses problems of excluding the relatively silent, less vocal, perhaps more timid speakers who are always in danger of being "overrun by more aggressive and articulate speakers." What must we know about the "narrative conventions for constructing individual and collective identities" (to say nothing of the issues) if we are to help mediator-facilitators listen more acutely, more sensitively, and nurture a richer public dialogue and negotiation as a result?

We can consider in addition a line of research questions that recent feminist scholarship raises about the issues of inclusion and exclusion. When mediator-facilitators of public disputes come to make judgments about the representation of "appropriate parties," they may well need to think about both formal rights or entitlements ("they live within the city boundaries!") and consequences and practical relationships ("pollution knows no jurisdictions"). In Carol Gilligan's terms, mediators face the challenges of integrating an ethics of justice ("Let's play by the rules; you'll get your turn in a minute") and an ethics of care ("What if they did that; what would the consequences be? Could you live with that?" Would anyone else be affected?) (Gilligan 1982, Kittay and Meyers 1987).

How might these mediators meet these challenges? A series of research questions are implied in Joan Tronto's acute discussion (1987:659–660) of a feminist ethics of care:

Advocates of an ethic of care face, as Gilligan puts it, "the moral problem of inclusion that hinges on the capacity to assume responsibility for care." It is easy to imagine that there will be some people or concerns about which we do not care. However, we might ask if our lack of care frees us from moral responsibility. . . . We do not care for everyone equally. We care more for those who are emotionally, physically, and even culturally closer to us. Thus an ethic of care could become a defense of caring only for one's own family, friends, group, nation. From this perspective, caring could become a justification for any set of conventional relationships. Any advocate of an ethic of care will need to address the questions, What are the appropriate boundaries of our caring? and more important, How far should the boundaries of caring be expanded?

Tronto's challenge to feminist ethics is a challenge to mediators and facilitators as well: Will the practicalities of distance, difference, and lack of familiarity lead to exclusivity, a defense of conventional problem definitions, a conservative if not reactionary limiting of public responsibility? How inclusive can and should mediator-facilitators be (Susskind 1981, Drake 1989)?

A series of research questions seems to follow then. How do public dispute-resolution practitioners frame these issues of representation and inclusiveness? How do they interpret their own roles: as neutrals, as "process specialists" with special claims to knowledge, as craftsperson-like practitioners with a trained practical judgment?

Tronto's challenge confronts educators and trainers too. Just what competence and judgment should mediator-facilitators of public disputes have? What senses of self, what interpretation of their own role, should be nurtured, encouraged, and which discouraged? We turn to these questions.

Toward a Research Agenda II: Taking Normative Political Theory Seriously or Having the Courage to Ask, "What Should We Want?"

Beyond Self-Paralysis: Reappropriating Moral Theory

A central difficulty here comes with the mediation community's own neglect of the language of moral theory. Reluctant to ascribe any but the most procedural responsibilities to mediators, students of the field invoke the language of neutrality, impartiality, and nonpartisanship without examining either the meaning of these concepts or the actualities of the practice they refer to very closely (Forester and Stitzel 1989, Young 1990).

The result is to produce research opportunities for students, surely, but also to reproduce a thin and emaciated conception of the ethical and political passions of mediation. Anyone who has ever mediated a dispute knows that hundreds of practical judgments must be made, but the labels of "neutrality" or "impartiality" hardly inform how those judgments *ought* to be made in diverse practical situations.

We need, in part, to understand and defend the sorts of commitments that mediators must make to the practice of dispute resolution and public deliberation. What commitments ought to inspire their practice? What commitments are simply necessary to doing what ought to be done? Assessing mediation as a social intervention practice, for example, Laue and Cormick (1978:221) argue against "neutrality" as an ethical principle to honor: "Since neutrality or claims to neutrality on the part of an intervenor in a community dispute almost always work to the advantage of the party in power, the intervenor should not claim to be (or worse, actually feel) neutral." We need further research—well-argued inquiry integrating evidence and purpose—that addresses what mediators in differing settings should feel, aspire to, and be committed to.

If we care to improve public dispute-resolution processes, we will not be able to avoid the question: Who should mediators be? We will need to face up to—that is, recognize, debate, and learn to nurture—the practical-ethical virtues required in exemplary public dispute-resolution practice. At issue here is our understanding not of the functions but of the *character* involved in working to mediate or facilitate public disputes.

The literature on these practical-ethical questions seems nonexistent. Not the justice, fairness, and distributive character of *outcomes* is at issue here but the *character* of the mediator's *practice:* gesture and word, attention and neglect, sensitivity and blindness, control and management, hope, wonder and frustration too. True, the field of public dispute resolution is still in its infancy. But the broader literature on mediation and dispute-resolution processes gives us little reason for hope. It reflects the prevailing (and professionally debilitating) blinders of American social science. It pretends for the most part that empirical questions can be addressed without presuming normative claims; thus it is largely descriptive and not prescriptive (Bernstein 1976, 1983, Schön 1983, Rorty 1982). Our social science sometimes tells us where we might go and how we might get there, but it all too rarely helps us to think about where we should go. To say that this disserves all those who must wrestle with the ambiguities of practice, all those in practice who are not merely robot-widget-makers mechanically following the rules, is to understate the case dramatically (Schön 1983).

This does not happen by accident. Social scientists are trained *not* to ask several of the questions we need most to ask if we are to understand the possibilities of public sector dispute resolution—if we are to understand not only who gets what and how they do, but what is at stake in the very enterprise.

We do need to know how current mediation processes work. We need to know how third parties work as dispute resolvers in public and private organizations. We must assess the promises and dangers of "informal justice" and of the privatization of dispute-resolution processes—and we are finding out (Kolb and Rubin 1989; Kressel et al. 1989).

But we are inquiring very little, and learning less, about what public sector dispute resolution, for example, should be. We are not asking this political and ethical question because we have been trained not to ask, and so we risk watching the drowning child as we do research on buoyancy.

Social scientists seek to explain and understand, so they test explanatory models and interpret social and cultural history. Nevertheless, although philosophers have long discredited the claim that "values" and "ends" cannot be discussed rationally—that we cannot learn about what we should want—still a simplistic doctrine of emotivism rules the social sciences, and the professions pay a steep price (Lyons 1984, MacIntyre 1981, Meer 1989, Nussbaum 1990b, Wallace 1988, Fay 1996). If we are to improve the world as well as interpret it, if we are to shape the study and practice of public deliberation and public dispute resolution, too, we must have the courage to ask, "Mediation and deliberative practice for what ends? What should we want, and why?"

We must have the courage to ask the normative, political, and ethical questions that reach beyond asking what we think about what we now observe, questions that also reach beyond "Who's left out?" or "What about rights?" or "What about the subtle biases here?" These are important questions, which thankfully are conventionally asked and whose answers matter. But although they are necessary for us, they are insufficient.

Taking Political Theory Seriously

We must also ask how, in particular cultural, historical, and institutional settings, public dispute-resolution processes should work: what qualities of participation mediators should encourage, what qualities of outcomes mediators should seek, what qualities of commitment, virtue, and judgment mediators should embody. These are the questions that political theorists, particularly those interested in democratic politics and practices, aspire to ask and address, if never to answer for all time.

Mediators and facilitators of public disputes are in part creators of political spaces, contingently collaborative, deliberative spaces in which disputants as citizens meet and seek to refashion their lived worlds. Along with asking how they have done what they have done last year, we should also ask, "Given whatever discretion we believe they have, how should such public dispute resolvers seek to create such political spaces—to what ends, with what stakes?"

Laue and Cormick (1978:229) put their view crisply, and it deserves careful debate, criticism, and refinement: "The basic responsibility of the intervenor, then, is to use skills, position, and power to further the empowerment of the powerless."

Susskind (1986:135) has distinguished the need to document dispute-resolution efforts from the need to evaluate those efforts: "Documentation means recording—from beginning to end—the expectations, impressions, and reflections of the participants. Evaluation means gauging the success of an effort once it is completed, according to explicit standards or criteria."

We need a good deal more: an ongoing debate, reflecting careful research, that will formulate and clarify those "standards and criteria" that we are to take as the very foundations of the evaluation efforts for which Susskind rightly calls. Just what is to be evaluated, what is to be asked, is too often *presumed* by social scientists and self-limitingly considered the sole province of (beware:) "theorists." Two rich examples of attempts at documentation and partial evaluation that help us ask what environmental dispute resolution should really be about are by Bingham (1986) and Susskind, Bacow, and Wheeler (1983).

These are questions not only to debate but to act on. Again, Hanna Pitkin (1984:289–290) is instructive:

The political theorist is not merely, or not exactly, a philosopher. Philosophy investigates those aspects of the human condition that could not be otherwise and that are so basic we are ordinarily not even aware of them. But politics concerns matters that might well be other than they are; it concerns the question "what shall we do?" Insofar as it directs itself toward matters that cannot be changed, it is misguided and will fail. Politics is the art of the possible. . . . Thus the political theorist . . . delineates, one might say, "what has to be accepted as given" from "what is to be done."

This delineation, Pitkin (pp. 290–291) tells us, is done with practical purposes in mind:

Change people's understandings of themselves and their human world, and they will change their conduct, and thereby that world. The political theorist is thus always a teacher as much as an observer or contemplator, and to the extent that his teaching succeeds, his subject matter will alter. So the distinguishing of necessities from possibilities is less like the drawing of a line than like a *Gestalt* switch: a reconceptualization of familiar details so that realities we feel we have always known suddenly become visible for the first time, familiar things suddenly take on a new aspect. . . .

Nothing is harder than to get people really to *see* what has always been before their eyes, particularly since such a changed vision will have implications for action (which may make uncomfortable demands on them) and interest (which may make it offensive to those who now hold privilege and power). Thus the political theorist faces a special problem of communication: in order to be understood, he must speak in terms familiar to his audience. . . . Yet he wants not to convey new information to them, but rather to change the terms, the conceptual framework through which they presently organize their information. . . . Political theorizing accordingly has its dangers, from ridicule to martyrdom.

Research on public sector dispute resolution may be, in these terms, information rich and concept poor, descriptively underway and prescriptively tied up at the dock. Accordingly, let us end by taking on the political theorist's task: to point to possibilities of seeing public dispute mediation practice anew—and risk not martyrdom but ridicule.

Neither Experts Nor Judges Nor Bureaucrats: Mediators as Critical Friends

Our working models or metaphors of the real practice of mediator-facilitators amount to pictures of empty boxes. Mediator-facilitators have no interests of their own, no commitments to anything but "the processes"! Typically they are to have a mirroring, but hardly felt, empathy perhaps, but certainly little sympathy. Feelings must be carefully controlled, if not altogether kept in a box within the box (cf. Friedman 1989:275–276).

Detailed ethnographic accounts of the actual practice of mediation seem quite few in number (Kolb 1994, Forester and Weiser 1996). The professional, quasi-how-to-do-it literature is filled with stages and steps, lists and hints, do's and don'ts, but all too little analysis of what it is to be a mediator (e.g., Moore 1986, Folberg and Taylor 1984). Academic research, even when not largely blinded disciplinarily, tends to be largely empiricist, whether focused on explanatory social-psychological accounts or more interpretive institutional ones—to the neglect of exploring the practical philosophical problem of articulating what sort of person, what sort of self, what sort of character mediator-facilitators *ought to seek to be*—in various historical settings, to be sure (cf., e.g., Kressel et al. 1989).

Consider what cares mediators should have. Presumably they must care that the rules of the game are honored. Seeking to establish relationships with the parties, perhaps they ought too to care for the dignity and voice of each particular party. Sullivan (1982:93) puts the problem in historical perspective:

> As early as the beginning of the nineteenth century, Alexis de Tocqueville noted the peculiarly American tendency to explain virtually all actions, especially those in public life, as the result of self-interest. . . . [Nevertheless] the disposition to treat others with dignity and respect is not simply a result of intelligent self-interest; rather, as Tocqueville acutely noted, respect for others is rooted in a deeper sense of mutuality and an habitual readiness to acknowledge worth in the other's point of view.

Can mediators afford not to respect the parties with whom they work? If they must respect them, are the character and demands of such respect not worth exploring (Forester 1982)? Then too, if mediators know that disputants will often come unexpectedly to see the issues before them in a new light, mustn't the mediators care to nurture that re-visioning, to resist and probe beneath initially articulated positions and interests, and so to midwife new understandings of problems and possibilities?

Mediators of public disputes must be close enough to listen but far enough away to manage the process. They must be sensitive enough to understand but be tough enough to ask hard questions. They must be attentive enough never to be dismissive, yet they must be insightful enough to probe for what may really matter. They must take each party seriously yet be able to laugh. They must be attuned to the varying and particular rationality of each party rather than presuming a stereotypical, perfectly clear and calculating economic rationality on each and every disputant's part. Managing a process of inquiry, deliberation, and negotiation, they must enable the parties to move ahead rather than tell them where to go.

In all these ways, mediators must be like people we already know. These people are not experts we consult for technical knowledge. They are not distant judges to whom we appeal to interpret the law. They are not neutral bureaucrats who tell us impassionately that the rules apply to everyone, and here's what it is that we must do.

Mediators must be instead more like respected, critical, and attentive friends—friends who can tell us when our clothes do not match, friends who can remind us when we are in danger of betraying ourselves, friends who can ask with us what is really possible, what we might shoot for, what we might live with.

In the deliberative process of democratic interaction, Barber (1984: 189) suggests, for example, "A neighbor is a stranger transformed by empathy and shared interests into a friend—an *artificial* friend, however, whose kinship is a contrivance of politics rather than natural or personal and private." Refining the point, Barber quotes MacIntyre (1981:146–147): "Friendship, of course, on Aristotle's view involves affection. But that affection arises within a relationship defined in terms of a common allegiance to and a common pursuit of goods. The affection is secondary." Notice too that friends may be among those we first turn to when we face conflicts *within ourselves*—when, significantly, not only our interests but our values and senses of self conflict.

This is emphatically not to say, in colloquial terms, that mediators can or must "befriend" the parties. This is not to say that mediators can or must develop affection for the parties. This is not to argue that mediators can or must develop prior relationships of deep attachment to the parties. Well, what then?

Aristotle (*Ethics:*1156b) distinguishes imperfect and perfect forms of friendship. Imperfect forms are those in which friends are sought to provide one with the useful or with pleasure; "those who wish for their friends' good for their friends' sake are friends in the truest sense, since their attitude is determined by what their friends are and not by incidental [opportunistic and contingent] considerations." A problem arises: If we wish mediators to seek disputants' good on the basis of the particularities of each disputant, who the disputants "are," are we in danger of privatizing the dispute-resolution process once more? Are we back to deal making? We are not, for the Aristotelian "disputant" is not a private individual but a member of a political world, a participant in a public sphere, a person whose good is bound up with their character of being in a political world with others (Nussbaum 1990b).

Perhaps we can come to think of mediators, and mediators can come to think of themselves, as more than business-like acquaintances yet less than affectionate friends: perhaps as new civic friends, "critical friends" who care enough to listen for more than what has been said, who care enough to wonder about what has been missed, who are engaged and collaborative enough to help, yet detached and independent enough to carry forward their own projects.

To take a mediator as a new, critical "civic friend" shares less with Fried's (1976) conception of friendship tied to the autonomy of the befriended than it does with Irwin's (1989) argument that friendship may promote, rather than subvert, justice. As Benhabib (1992:11) argues, "In the democratic polity, the gap between the demands of justice, as these articulate principles of moral right, and the demands of virtue, as this defines the quality of our relations to others in the lifeworld, can be bridged by cultivating qualities of civic friendship and solidarity." Here we bridge traditions of Kant and Aristotle, Habermas and Nussbaum, through Benhabib's careful attention to democratic conversation and democratic character too.

The point here is not to presume a determinate concept of friendship and its virtues but instead to motivate research to explore the embodied possibilities suggested by the metaphor. In particular we should explore gendered (and androgynous) conceptions of friendship (Raymond 1986). Highlighting the nonaffective dimensions of friendship, Friedman (1989: 287), for example, notes that "friendship is more likely than many other close personal relationships to provide social support for people who are idiosyncratic, whose unconventional values and deviant life-styles make them victims of intolerance from family members and others who are unwillingly related to them. In this regard, friendship has socially disruptive possibilities, for out of the unconventional living which it helps to sustain there often arise influential forces for social change."

Sandel suggests that were rationally calculating, clearly self-knowing autonomous individuals to exist, much of what we understand to be the character of friendship, and the character of friends, would be altogether irrelevant to us. But it is not, and we may surmise, for good reason. Sandel (1982:181) writes, "Where seeking my good is bound up with exploring my identity and interpreting my life history, the knowledge I seek is less transparent to me and less opaque to others. Friendship becomes a way of knowing as well as liking. Uncertain which path to take, I consult a friend who knows me well, and together we deliberate, offering and assessing by turns competing descriptions of the person I am, and of the alternatives I face as they bear on my identity. To take seriously such deliberation is to allow that my friend may grasp something I have missed, may offer a more adequate account of the way my identity is engaged in the alternatives before me."

Less, then, like experts, judges, or implausibly neutral bureaucrats, mediators of public disputes should be seen as new, civic friends in the making: new friends of a diverse public; new friends who hope to seek out those affected and who will attend to their inclusion; new, civic friends who can create a space for speaking and listening, for difference and respect, for the joint search for new possibilities, and ultimately for newly fashioned agreements about how we shall live together.

IV

Participatory Planning Can Transform Public Disputes

7

On Not Leaving Your Pain at the Door: Political Deliberation, Critical Pragmatism, and Traumatic Histories

[On writing Holocaust history] It is important to keep in mind that one can acknowledge the *fact* of an event, that is, that it happened, and yet continue to disavow the traumatizing impact of the same event.

—Eric Santner, 1992

[Environmentalist to a state land use planner after a contentious meeting] I'm sorry that we're so critical, but we've been burned.

—John Forester and Brian Kreiswirth, 1993b

Practical discourse is not a procedure for generating justified norms but a procedure for testing the validity of norms that are being proposed and hypothetically considered for adoption.

—Jürgen Habermas, 1990

Three Challenges of Deliberative Learning: Means, Ends, and Identities Too

Democracy can be painful, and any theory of political participation that obscures this fact should make us suspicious. To do greater justice to the agonies and possibilities of real deliberative practices, critical social theory must take into account, much more clearly than it does, the painful histories that citizens bring to many public deliberations. Because many accounts of political deliberation are too antiseptic, too free of the historical legacies of pain and suffering, racism and displacement, that citizens bring to decision-making arenas, we need to imagine and conduct political deliberations in new ways (Michelman 1988, Barber 1984, Majone 1989, Manin 1987, Bryson and Crosby 1992).

We can approach this problem in four parts. First, because Habermas's work appears at times to be not just provocative but also disturbingly ahistorical and overidealized, we take a fresh approach: we subordinate his more philosophical discourse ethics to his more sociological analysis of communicative action. Second, we consider a striking and vivid case of participatory action research that shows how deliberation can involve not only justifying claims, but working through past loss as well. Third, we consider lessons drawn from historians and psychiatrists concerned with the representation and working through of trauma. Fourth, and finally, we draw implications for political deliberation and its ritual support, and we end with an instructive example of deliberative community planning.

In daily practice, information is power. Planners and policy analysts regularly and selectively shape what parties know or believe about cases, how they defer or consent to norms, and how they develop or lose trust in the identities of others (Forester 1989, 1991b, 1993b; Innes 1995b; Healey 1997). When citizens participate in actual deliberations, they too face three similar challenges: to learn about strategies that will or will not work, to learn about responsibilities and obligations as they assess proposed norms of action, and to learn about themselves and the others with whom they might act. In political deliberations, accordingly, we must learn about others and ourselves, too, in addition to what is challenging and difficult enough: learning about means and ends.

Consider for a moment the problems of learning about means and ends. We have substantial technical knowledge about probing means and strategies to reach objectives, but we know much less about probing ends. Many wonder, for example, if the probing of ends is even a rational enterprise (Nussbaum 1990a). Can we really learn about "what we should want"? Can we learn about "value" at all? Notice the dangers here: if we craft a process helping "stakeholders" to learn about their ends, we could easily be tempted to take their initially valued, even cherished, "ends" a bit less seriously—for, after all, once they have been through our dialogical boot camp, they'll *really* know what they want!

This must be a virulent form of the planner's or politician's temptation, but it points to deep problems in democratic theory and practice. How are we to accord respect, not a condescending tolerance, and listen seriously to those who seem to be demanding and righteous fist pounders—those whose sense of ends we fully expect to change with the experience and information, the education and transformative magic, of democratic practice? How are we to structure and encourage democratic deliberations in which participants can learn so much that their senses of hope and possibility, relevance and significance, interests and priorities shift? If we know, as Jon Elster vividly suggests, that "adaptive preferences" reflecting political resignation are commonplace, how should we design processes of participation and voice, accountability and review, to shape, inform, or educate preferences?[1]

These difficulties with deliberations about ends follow from the basic insight that we can and do learn about value, and not just articulated values, all the time. But the challenges of deliberation do not stop here, for as deliberators we hope also to learn about ourselves and others—trying to move beyond "us and them," we try to imagine what new relationships we might build. We hope to see others as more than just "greens" or "reds" or "blacks" or "labor" or "management." Listen to the evaluators of recent environmental risk analysis projects in six states:

> Every project has shown that the first emotion to arise between technical committees and policy or public advisory committees is likely to be distrust. The scientists fear that the policy-makers will disregard their data and make judgments based primarily on emotion, self-interest, or political expediency. The policy makers assume that the technical committee members will be either incompetent bureaucrats or dishonest manipulators trying to justify ever bigger expenditures for their own programs. . . . The resulting distrust . . . can undermine implementation efforts. (Minard, Jones, and Patterson 1993:12)

We need to explore how practical deliberators might listen to others' claims and understand how they are shaped by institutional and cultural histories. We need to understand others' and our own emotions of fear and suspicion, anxiety and resentment, compassion and generosity not simply as brute facts but, in Martha Nussbaum's words, as "modes of vision, or recognition," as avenues for learning rather than as necessary evils (Nussbaum 1990a).[2]

Habermas the Critical Pragmatist versus Habermas the Theorist of Justice

Habermas's discourse ethics appears to provide a model of public deliberation, but it really does not do so, and its role is quite narrow, even if deep. To understand participatory processes better, we should extend not Habermas's moral theory but his sociology of action—his concern with the precariousness, institutional contingencies, and political vulnerabilities of ordinary understanding and interaction, our abilities to interpret and "make sense together." This interpretive strategy—reading Habermas sociologically as a critical pragmatist rather than more narrowly as a Kantian moral theorist—can make Habermas's work far more useful than many believe (Flyvbjerg 1998, Forester 1993b).

Habermas's discourse ethics has limited prescriptive use; he never intended it to be a how-to manual. But his account of communicative, performative action leads to many fruitful questions of research and practical politics. In particular, this pragmatic reading of Habermas illuminates not only issues of misinformation, deceit, obfuscation, and abuse of power, illustrated profusely in Flyvbjerg's own case study of transport planning in Aalborg (1998), for example, but also issues of the construction and reconstruction of political identity, our political sense of self, in deliberative processes (White 1985, O'Neill 1974, Sager 1994a).

Habermas sees a narrow but important role for "discourse ethics": to examine the normative validity of public action-guiding norms, to examine not just whether all affected participants might *accept* a norm, but whether the norm *deserves to be accepted by them, given the process in which they might consider it.* Were the people fully informed? Were better alternatives suppressed? Was their acceptance gained through intimidation? His concern is with justice and not the good, with our understanding of normative validity and *not* with regulating value preferences themselves:

> While cultural values may imply a claim to intersubjective acceptance, they are so inextricably intertwined with the totality of a particular form of life that they cannot be said to claim normative validity in the strict sense. By their very nature, cultural values are at best *candidates* for embodiment in norms that are designed to express a general interest. . . . A de-ontological ethics . . . deals *not with value preferences but with the normative validity of norms of action.* (1990, p. 104, latter emphasis added)[3]

Critics of Habermas badly misunderstand his work when they treat his procedural model of practical discourse as a method for organizing real deliberations. He proposes an in-principle model to test, not to generate, good policy, and he is quite explicit:

> Discourse ethics . . . provides no substantive guidelines but only a procedure: practical discourse. Practical discourse is *not a procedure for generating justified norms* but a procedure *for testing the validity* of norms that are being proposed and hypothetically considered for adoption. That means that practical discourses depend on content brought to them from outside. . . . Practical discourses are always related to the concrete point of departure of a disturbed normative agreement. These antecedent disruptions determine the topics that are up for discussion. (Habermas, 1990, p. 103, emphasis added)

Among these "disturbed normative agreements" that can generate "the topics that are up for discussion," we can find the systematic, policy-sanctioned expropriation of one's land, the pervasive racism threatening the lives and prospects of one's family members, the disregard of law enforcement officials in the face of sexual violence, the licensing of industrial interests to destroy precious natural resources, and so on. But these "antecedent disruptions" do more than generate topics for ethical debate; they also produce pain, humiliation, outrage, feelings of betrayal and desires for revenge, trauma and the after-effects of trauma. These disruptions lead—through social movements or individuals—to claims on past and existing authorities and future generations too. Habermas takes these claims of the oppressed, victims, and disenfranchised as prior to, and feeding into, real public deliberation. He suggests that we could hardly imagine a "discourse ethics" testing the validity of proposed norms without considering those claims.

Nevertheless, Habermas's preoccupation with the justice of policies—with the possible validity of norms—can distract us from the prior challenge of learning about other citizens, including oppressed, victimized, traumatized, or suffering people, whose claims we can understand only if we recognize their historical backgrounds and entanglements, their historically rooted aspirations, and their interpretations of needs and senses of self.[4]

So we have two aspects of critical social theory in tension with each other.[5] The sociological, critical-pragmatist Habermas links real communicative interaction to the always contingent and iffy formation of social identity, while the more philosophical Habermas articulates a "discourse ethics" that presumes, but does not itself explore, such identity.[6] This tension can be constructive if, and only if, we resist the temptation to imagine real political deliberations as applied "discourse ethics." Again, testing a justification is not generating it. Habermas teaches us about the former, not the latter. Readers of Habermas have assumed too easily that analyzing the conditions of justification tells us how to achieve justice. But it does not—any more than the analysis of Michael Jordan's moves tells us how to move like Michael Jordan!

Our problem now looks something like this. Public deliberations bring parties face to face across lines of class, race, gender, and territory—parties who bring histories of being doers and done-to, histories of rights enjoyed and rights betrayed, histories of hopes deferred and pains suffered. Hardly all knowing, these parties bring values and preferences, senses of self and needs, that can change as political possibilities change.[7] Their priorities can shift as they learn about cases and learn, more significantly, about one another too. When I discover, for example, that your priorities are not what I thought, my priorities might change too, especially if we are interdependent and have to come to terms with each other. When I discover that you suffer in ways I did not see, I may understand better not just your words and not just your needs, but our possible future relations—and so my future interests too.[8] Learning about each other and the issues at hand too, deliberating parties can create public value: from the value of mutual recognition to that of their empowered capacities to act, singly or together (cf. Moore 1995, Susskind and Field 1996, Forester 1999b).

Seyla Benhabib argues that Habermas's approach to discourse ethics puts aside too many issues of autonomy and interdependence, the "moral texture" of interrelatedness, entanglement, and attachment that makes us who we are (Benhabib and Dallmayr 1990:356–357):[9]

> This rationalist bias of universalist theories in the Kantian tradition has at least two consequences: first, by ignoring or rather by abstracting away from the embedded, contingent, and finite aspects of human beings, these theories are blind to the variety and richness as well as significance of emotional and moral development. . . .
>
> Second, the neglect of the contingent beginnings of moral personality and character also leads to a distorted vision of certain human relationships and of their *moral texture,* precisely because universalist and proceduralist ethical theorists confuse the moral ideal of autonomy with the vision of the self "as a mushroom" (Hobbes). . . . Current constructions of the "moral point of view" so lopsidedly privilege either the *homo economicus* or the *homo politicus* that they exclude all familial and other personal relations of dependence from their purview. . . . [But]

moral autonomy can also be understood as growth and change, sustained by a network of relationships.

If we wish to envision public deliberations in which citizens will not just argue about norms or strategies but also recognize the texture and often painful history of their social relatedness, we should appropriate Habermas's work as a "critical pragmatism"—"critical" because concerned with ethics and justification, a "pragmatism" because concerned with practical action, history and change.[10] Presuming no equality of power and instead probing the precariousness of intersubjective understandings and agreements, a pragmatic Habermasian analysis of deliberation leads us directly to questions of power and hegemony, agenda setting and the contestable reproduction of citizens' knowledge, consent, and social relationships (Forester 1993b, 1989).

This broader view of public deliberation illuminates proposals like Robert Reich's (1988; cf. chapter 6 above) to assess public policy analysis as it fosters deliberative processes of "civic discovery," and Benjamin Barber's (1984) to encourage a "strong democracy" in which parties not only protect their autonomy but learn with one another, and learn how they can act together as well.

Imagining deliberation to involve questions of strategies, goals, and identities too in political settings of structured inequality, consider now a case of political, if not quite public, deliberation in which the valuable evolution of identity and the significant recognition and partial working through of trauma figured prominently.

Deliberation in the Shadow of Trauma

A Cornell researcher-activist, Mary Jo Dudley, returned to Colombia recently to continue her studies of weakly organized, isolated informal sector workers. She had several contacts with a loosely organized group of domestic workers who met periodically in workshops on legal rights and in subgroups concerned with public education about the low pay and few protections these workers enjoyed.

Mary Jo had had years of experience with video and television production. When she asked to meet with the domestic workers' organization in some regular way, she was asked, somewhat skeptically, "And what can you do for us?" She answered that she could train their education subgroup to use video technology if they wished (Dudley 1996).

Mary Jo's story of the women's use of the video equipment is fascinating, not because of any resulting video production, but in spite of it. Beginning with simple interview exercises done by the domestic workers with one another as they learned to hold and use the video equipment, Mary Jo watched a process develop in which the workers were soon telling one another stories of sexual harassment and abuse at work that they had never told others before. "I've never told even my sister, or my mother, this story," one and another would say, and the group found itself speaking and listening in ways it had never done before.

But more happened too. Angered by another video about domestic workers done sympathetically but nevertheless in a caricatured way, the workers wanted to document not only their experiences but the views of the employers they worked for and then too the views of the broader public. After sharing their own stories with one another, these women felt strongly enough to go to public parks to begin to interview men and women at large about *their perceptions* of domestic workers—their status and situation.

To questions about harassment at work, for example, they heard comments all too familiar in the United States: roughly, "Oh, this depends. Many times this is a problem with the way the woman dresses, with her character, with her lack of courage to resist . . . ," and so on.

In a presentation about this process of self-styled participatory action research, Mary Jo presented several problems for us to consider: Just what had happened in the process? Originally intending to do a video to document their experience, the domestic workers changed their strategy to document the views of others and to use the video camera to engage in a public dialogue about the importance of domestic work in the Colombian context. In these videotaped interviews in public parks, Mary Jo was surprised when the women's general questions became more pointed, raising questions about general mistreatment and, in particular, sexual harassment and abuse that were all too invisible to the public. Originally participating in periodic workshops with little opportunity to learn about each other, the workers now found themselves telling one another stories they had never told their closest family members and so building relationships of solidarity and trust they had not anticipated. That solidarity and trust encouraged these women to develop public voices in ways they had not intended. Originally thinking about presenting their own experiences, the workers began now to analyze the views of their multiple audiences and ask what arguments and issues they could most compellingly present: linking their own desires for education, for example, to the employers' desires that the workers help the employers' children with their school work.

The process of deliberation, Mary Jo reported, became much more important than the product. As these workers reviewed the videotapes they had produced, they listened carefully once again to one another's experiences. They listened now for statements they could select to represent themselves as a group in the final video. But the process was not simply a product-driven one, for the result of the process was *not* a successfully completed video production as originally intended, but a transformed organization of domestic workers—transformed because they had changed their perceptions of each other, deepened their relationships and commitments to one another, supported one another in listening to the traumas they had endured, strengthened one another enough to take their case face to face in public arenas, learned more about the audiences to whom they hoped to take their case, and changed their sense of political strategy along the way.

What might we learn from Mary Jo's experience about the broader possibilities of public discourse and learning? We should ask, in particular, what place the storytelling of trauma might have in democratic deliberations.

Consider several points. First, Mary Jo's function was less instrumental, even though she introduced the video technology and accompanying techniques to the group, than it seemed catalytic. If we evaluated her role as a purely instrumental one designed to promote deliberation through the production of a video, we would find little to measure. We would be like the mythical drunk looking for his keys under the lamppost because that was where the light was, even though he had dropped them half a block away.

Second, the group members learned from one another less it seemed through political arguments that established claims of legitimate entitlements and dignity than through quite intimate and powerful storytelling. As chapter 5 suggested, the significance of the video technology that Mary Jo introduced to the group may have come far more in its enabling of rituals of recognition, safe spaces of storytelling, and mutual response than in its technological power to produce images.

Third, the domestic workers' education group seemed to learn in ways that changed themselves, their senses of one another, and their commitments to public action. Their senses of priority and significance seemed to change as they developed closer ties to one another, envisioned new possibilities of public action, and imagined the felt needs of others to which their arguments might appeal.

Fourth, their storytelling rituals allowed the workers to begin to work through their experiences of trauma and abuse in ways they had not been able to do before. Somehow these deliberative encounters that probed personal experiences in a protected space allowed the workers not only to verbalize painful experiences, but to recognize the similar experiences of others, and thereby to become less isolated and less alone, to support one another, to recognize one another's pain and vulnerability. Judith Herman's analysis of trauma and the work of recovery is instructive here. She writes, "Because the traumatic syndromes have basic features in common, the recovery process also follows a common pathway. The fundamental stages of recovery are establishing safety, reconstructing the trauma story, and restoring the connection between survivors and their community" (Herman 1992:3). Writing of the work of therapy groups devoted to that second stage of remembrance and mourning, reconstructing the trauma story, Herman indicates what the group's members can achieve: "The support of the group enables individuals to take emotional risks beyond what they had believed to be the limits of their capability. The examples of individual courage and success inspire a group with optimism and hope, even as the group is immersed in horror and grief" (p. 223).

Fifth, their own personal and intimate stories fit and began to do justice to their experiences in ways that more general political arguments (in workshops on legal rights, for example) had not. Further, the sharing of these traumas strengthened the group, perhaps because of the reciprocity enacted in the mutual work of telling and listening carefully, sensitively, and respectfully.[11] In her powerful study of survivors of trauma, Judith Herman (1992) suggests the potentials of survivor groups: "The solidarity of a group provides the strongest protection against terror and despair, and the strongest antidote to traumatic experience. Trauma isolates; the group re-creates a sense of belonging. Trauma shames and stigmatizes; the group bears witness and affirms. . . . Trauma dehumanizes the victim; the group restores her humanity" (p. 214).

We can now see, sixth, that deliberation in the shadow of trauma may require much more than a neutral political space in which to debate claims. The domestic workers' case suggests that arguments about strategy were interwoven with elements of the working through of traumatic experience: the sensitive representation rather than dismissal of the workers' victimization, changes in the understanding of self and other, development of the courage to speak publicly, and changes in the workers' senses of significance and preference, issues, and goals. At the same time, as Mary Jo notes, "The trauma has many levels, and if society blames the victim, the possibility of creating the safe space prerequisite for healing is much more difficult" (personal communication, May 27, 1998).

But should we think that learning of this kind is relevant only to relatively homogeneous groups that share a common history of disruptive, traumatic experiences? Or will it be still *more* important in more pluralistic settings to begin to acknowledge (rather than continue to ignore or suppress) such identity-shaping traumas, if parties in those deliberations are not to feel ignored, dismissed, or even complicit in their own invisibility and silence?

One problem that arises here is this: What is the relationship between the working through of historical traumas (of racism, exploitation, expropriation, displacement, and murder), on the one hand, and the processes of public deliberations, on the other? Surely it will not do to restrict our accounts of political deliberations to encounters in which participants have already come fully to terms with the legacies of victimization, resignation, silence, and humiliation they have inherited. To sanitize deliberations in this way would make a rehomogenized mockery of any supposed pluralist character they might have. We should explore instead how the working through of historical trauma and processes of political deliberation might coexist, if not be closely intertwined.

These are distinctive challenges, irreducible to, yet shaping one another. Structures of deliberation can encourage or displace processes of acknowledging and working through collective suffering. In turn, the sources that might value such working through—including social movements articulating racial, gendered, and even territorial issues—might encourage or subvert public deliberations. Think here of the history of feminist consciousness-raising groups, ethnic empowerment efforts, and liberation movements more generally.

Nevertheless, current accounts of deliberative democracy seem strangely silent about the place of historical trauma and victimization, pain and suffering, as elements of deliberative consideration.[12] Clearly this is no oversight, for theorists of democratic deliberation are committed to the inclusion within political deliberations not only of pained voices but of painful claims themselves.[13] But we can write about fair procedures far more easily, it seems, than we can about the open wounds of our history.

We need not only to make the case for inclusion but to explore the real challenges of inclusion in practice. Until our political analyses and theories do that, we will continue to hear the sound of one hand clapping: deconstructive critiques without reconstructive political analyses. Fortunately, we can learn much here from two distinct but related sources: historians writing about the representation of trauma in Holocaust historiography and psychiatrists writing about trauma and recovery.

Writing History and Re-presenting Trauma

Historians' work suggests that if we wish, as deliberative democrats, to begin to explore processes of the working through of suffering, we must consider how we are to represent, articulate, and try to recognize traumatic history and traumatic legacies in the first place.

Saul Friedlander (1992), for example, begins with Freud's definition of trauma and explores the historian's tasks and the problem of working through:

In *Beyond the Pleasure Principle,* Freud defines as traumatic "any excitations from outside which are powerful enough to break through the protective shield (*Reizschutz*). . . . Such an event as an external trauma is bound to provoke a disturbance on a large scale in the functioning of the organism's energy and to set in motion every possible defense measure. . . .

Aside from being aware and trying to overcome . . . defenses . . . [of denial, splitting off, and fragmentation], the major difficulty of historians of the *Shoah,* when confronted with echoes of the traumatic past, is to keep some measure of balance between the emotion recurrently breaking through the "protective shield" and numbness that protects this very shield. In fact, the numbing or distancing effect of intellectual work on the *Shoah* is unavoidable and necessary; the recurrence of strong emotional impact is also often unforeseeable and necessary.

Working-through means, first, being aware of both tendencies, allowing for a measure of balance between the two whenever possible. But neither the protective numbing nor the disruptive emotion is entirely accessible to consciousness. . . . A main aspect of working-through, however, lies elsewhere: it entails, for the historian, the imperative of rendering as truthful an account as documents and testimonials will allow, *without giving in to the temptation of closure.* Closure in this case would represent an obvious avoidance of what remains indeterminate, elusive and opaque. (pp. 51–52)

Friedlander reminds us that the representation—literally the re-presentation—of historical trauma presents real emotional dangers for the historian and the psychologist. Little wonder, then, that we find historians worrying about responses of avoidance, denial, premature closure, abstraction, and resignation.[14] We should hardly be surprised if similar psychic dangers and the temptations of evasion and denial confront our public officials, planners, and citizens meeting in acrimonious public deliberations.

Friedlander (1992) understands working through to work against the terms of bureaucratic accounting and political simplification:

Working-through means confronting the individual voice in a field dominated by political decisions and administrative decrees which neutralize the concreteness of despair and death. The *Alltagsgeschichte* [the history of everyday life] of German society has its necessary shadow: the *Alltagsgeschichte* of the victims. . . . Working-through does mean a confrontation with the starkest factual information which loses its weight when merely taken as data. Raul Hilberg mentions a report sent by the German military headquarters in the Black Sea port of Mariupol in 1941. In only a single line it stated that "8,000 Jews were executed by the Security Service." Working-through ultimately means testing the limits of necessary and ever-defeated imagination. (pp. 54–55)

Friedlander asks us to avoid the extremes of both supersensitivity, with its resulting paralysis, and reductive dismissal. He asks us to open ourselves up, to listen closely, but to protect ourselves at the same time.

In a related discussion of the representation of trauma in the writing of history, Eric Santner (1992) contrasts the stances of addressing trauma through critical work of mourning or through a defensive and evasive "narrative fetishism":

> The work of mourning is a process of elaborating and integrating the reality of loss or traumatic shock by remembering and repeating it in symbolically and dialogically mediated doses. . . . Narrative fetishism, by contrast, is the way an inability or refusal to mourn emplots traumatic events; it is a strategy of undoing, in fantasy, the need for mourning by simulating a condition of intactness, typically by situating the site and origin of loss elsewhere. . . .
>
> Both mourning and narrative fetishism . . . are strategies whereby groups and individuals reconstruct their vitality and identity in the wake of trauma. The crucial difference between the two modes of repair has to do with the willingness or capacity to include the traumatic event in one's efforts to reformulate and reconstitute identity. (pp. 144, 152)

There may well be parallel arguments to make regarding the silences of deliberative democrats in the face of racism, expropriation, sexual violence, and more. The question here, simplified perhaps, is this: What kind of talk about loss is democratic talk? This is a question not of the epistemology of democratic debate and discourse, but of its rhetoric—its tone, sensitivity, and character.[15] But if Friedlander and Santner signal dangers to us, Dominic LaCapra suggests that the performance of representing historical trauma can itself shape subsequent processes of acting out and working through. At first echoing Friedlander and Santner, LaCapra (1994) writes,

> In different ways in various disciplines or areas of discourse and representation, the *Shoah* calls for a response that does not deny its traumatic nature or cover it over through a fetishistic or redemptive narrative that makes believe it did not occur or compensates too readily for it. In one sense, what is necessary is a discourse of trauma that itself undergoes—and indicates that one undergoes—a process of at least muted trauma insofar as one has tried to understand events and empathize with victims. (p. 220)

But then LaCapra poses two questions that indicate the alternatively recuperative or reductive possibilities of the historian's own representation of trauma. He asks,

> First, does modern society have suitable public rituals that would help one come to terms with melancholia and engage in possibly regenerative processes of mourning, even if in extremely traumatic cases an idealized notion of full recovery may be misleading? Second, who it is that one mourns and how can one specify

the object of mourning in ways that are both ethico-politically desirable and effective in reducing anxiety to tolerable limits? (pp. 213–214)

LaCapra poses the political and psychological problems of mourning in public rather than in more narrowly individualistic terms:[16] "One may ask whether the social space of the empathic witness is sufficiently occupied by a one-to-one relationship or whether it requires—for any publicly effective (and perhaps for any individually durable) process of mourning—the role of more widely instituted practices" (p. 215).

Noting that "any process of mourning . . . would require some degree of specification of the object, which anxiety would always exceed," LaCapra suggests a form of our representation of trauma that itself has ritualized qualities:

Insofar as historiographic *discourse could itself validly have a ritual dimension* without sacrificing its critical nature, it too might assist in some small way in *facilitating warranted public processes of mourning*. At the very least, one might point out that *the idea of an appropriate language*—indeed, an acceptable rhythm between language and silence—in attempting to render certain phenomena *depends on ritual as well as aesthetic criteria*. (p. 215, emphasis added)

It is crucial here to appreciate that LaCapra is linking our representation and means of acknowledgment of past suffering to public processes of mourning, working through, and going forward. This is a problem we all face, personally and publicly as well.

I quote at length here not to confuse esoteric debates of historians with pragmatic tasks of deliberative democrats, but instead to build on the historians' insights so we might improve actual deliberative encounters. The historians are warning us that the ample representation of trauma is akin to unleashing hell; that the suppression of the pain of racism may be not only evasion and more racism, but also in part a matter of self-protection. Both the prospects of anxiety and acting out *and* the work of denial and facile closure threaten victims, and those who listen to them. And deliberative democrats, nothing and no one if they are not listeners, face such problems all the time. The point is certainly *not* to excuse evasion and denial, but to interpret them in order to change them.

But the historians teach us more here too: that unless we understand more clearly the relationships between our re-presentation of trauma and public processes of mourning and identity re-formation, we are unlikely to sustain meaningful public deliberations in the shadow of trauma, massive poverty and suffering, and expropriation (Marris 1975, 1986). To put this more institutionally, the historians suggest, as I have argued in chapter 5, that we must understand much better than we do the ritualized aspects of public deliberations in which citizens can tell their stories and make their arguments, in which they can begin to recognize one another's experiences and suffering, their senses of need and aspirations too.[17]

Trauma and Tentative, Pragmatic Recovery Processes

Just as historians like LaCapra link the re-presentation of trauma to the facilitation of ritual processes that might enable the beginnings of recuperative working through to progress rather than remain repressed, so can the insights of psychiatrists who have worked with trauma survivors help us to understand processes of working-through. Judith Herman (1992), for example, writes of both the promises and dangers of group work for survivors of war and sexual abuse:

> Because traumatized people feel so alienated by their experience, survivor groups have a special place in the recovery process. Such groups afford a degree of support and understanding . . . not available in the survivor's ordinary social environment. The encounter with others who have undergone similar trials dissolves feelings of isolation, shame, and stigma. . . . Groups provide the possibility not only of mutually rewarding relationships but also of collective empowerment. (pp. 215–216)

But Herman acknowledges the complexities of forming such groups:

> While in principle groups for survivors are a good idea, in practice . . . to organize a successful group is no simple matter. . . . The destructive potential of groups is equal to their therapeutic promise. The role of the group leader carries with it a risk of the irresponsible exercise of authority. Conflicts that erupt among group members can all too easily re-create the dynamics of the traumatic event, with group members assuming the roles of perpetrator, accomplice, bystander, victim, and rescuer. . . . To be successful, a group must have a clear and focused understanding of its therapeutic task and a structure that protects all participants adequately against the dangers of traumatic reenactment. Though groups may vary widely in composition and structure, these basic conditions must be fulfilled without exception. (p. 217)

We would be mistaken here to think all this relevant for so-called survivor groups but irrelevant for public deliberations. Those deliberative encounters are not survivor groups, but they can exclude survivors only at the risk of no longer being "public."

Representing (versus Suppressing) Loss in Public Deliberations Is a Practical Matter

If our public deliberations are to include those who have lost family members to ghetto violence, lost land to developers or the state, lost life chances again and again in the face of racism and sexism, blindness to those citizens' needs for recognition and support will express not neutrality but callous disregard of their distinctive histories and identities. Such blindness promises to produce not fairness but a charade in which citizens fail to learn from one another, fail to appreciate one another's histories, fail to recognize common vulnerabilities and aspirations, and fail to craft policies and actions responding to their real needs rather than to those of stereotypes.

This argument suggests a challenge of political design informed by history, anthropology, and psychiatry no less than by deontological or procedural ethics: we must be able to design deliberative processes and rituals that do not retraumatize citizens (as adversarial public hearings or courtroom procedures might) but instead attend systematically to citizens' needs for safety, voice, recognition, and then public interaction and action together.[18]

To make matters more difficult here, though, we should remember that public deliberations not only may bring together survivors of many kinds, but bring them together with perpetrators or at least those supporting them.[19] This may take one form in a court of law, another in cases of ethnic and racial tensions, and yet another in public hearings in which the environmentally polluted face the polluters, in which long-time community residents face gentrifying developers, in which urban development interests face rural conservationists, and so on (Richards 1996).

I have argued that public deliberations can exclude the legacy of trauma and identity-shaping suffering only at the cost of no longer being either public or deliberative. It would be mistaken to take this argument as showing the flimsiness of procedural accounts of discourse ethics. Quite the contrary, the challenge to restructure deliberative processes to even begin to acknowledge, mourn, and work through unspeakable loss means to develop new rituals of public discourse—new rituals requiring assurances of safety and protection, shared processes of voice and recognition, articulation and participation (Healey and Hillier 1996, Baum 1990, Marris 1975).[20]

Such newly articulated rituals of political and public deliberation will not be "one size fits all" techniques. But they are likely to share common characteristics. They will create safe spaces for participation, both for homogeneous groups as they articulate and rearticulate needs, and for heterogeneous groups facing each other, listening to and making claims on each other. They will create spaces for the telling of the political stories that characterize not only the issues at hand but the parties themselves. They will honor virtues of listening, attentiveness, and recognition, respect for differences and for rules protecting parties as well. They will require convenors—planners, public officials, community leaders—who can appreciate affect and attachment as desirable goods, not necessary evils. They will require leadership, from convenors and parties alike, not with facile public relations skills but with engaged courage and detached strength, with the ability to recognize pain and point toward the future, with the ability to imagine and explore relationships and actions no one thought possible before their deliberation together (Reardon 1993, Forester, Pitt, and Welsh 1993, chapter 6 above, Kirp et al. 1989).

Here we can again appreciate the importance of Habermas's work, not prescriptively, not as moral theory, but as a critical pragmatic analysis of speech and action. Habermas can help us to understand better political, communicative interaction—its staging and vulnerabilities, its precarious and socially constructed character, the ways power operates in it through the reproduction of community members' beliefs, consent, identities, and attention—and the stakes involved.

Neither a single vision of the good life nor procedures to test the justice of proposed policies will do here. In the shadow of traumatic histories, public deliberations will have to enable identity formation as well as strategy testing, acknowledgment and mourning of loss as well as the articulation and reinterpretation of ends. When we learn about the significant historical experiences of others and articulate our own in public settings, we may change ourselves as well as our strategies and senses of priorities. But such transformations of self-understanding, other-understanding, and our resulting relationships will be too easily marginalized if we reduce our deliberative focus to means and ends alone. We need to learn about strategies, about norms, and about selves, transforming all three as we go.

We cannot turn public deliberations into therapy sessions, but neither can we ask citizens to leave their painful histories at the door. How we should engage those histories and structure our deliberations to do it are issues that deliberative democrats need to address.

To explore these questions, we will have to add to our current understanding of scientific and normative debate and discourse a far more clearly articulated sense than we now have of the often painfully limited processes of recognition and working through. We will also have to learn to write and speak in the face of loss far more sensitively and imaginatively, far more practically, than we now do.[21]

A Glimpse of Possibilities: Participatory Action Research in Community Planning

Consider finally an account of a housing-related community planning case that provides more than a glimpse of such deliberation and learning. The case involves not just issues of housing ownership, but questions of race and suspicion, class divisions and stakeholders' presumptions, planners' learning and collaborative action. The narrator is Marie Kennedy, professor and community planner at the University of Massachusetts, Boston:

When there are issues around fundamental values that I hold—where there is contention—I don't just clam up and say, "We're not going to agree on that and that's going to cause a problem, so we won't talk about it. We'll just work on this other thing over here that everybody can agree on."

I don't agree with that because I feel that's disrespectful and patronizing. It's like saying that I'm over here and I've got my good politics, and those folks over there are just too narrow-minded and backward to understand and to be able to have an intelligent discussion about this.

So I'll bring it up and we'll have a discussion, and we may have a fight, and they may ask me to leave. It's never happened, but it's come close to happening in twenty-some years of doing these projects.

Actually, that's a criticism I have of some organizing approaches: taking the people where they're at and then assuming that somehow you can move through certain issues and then you can move onto more complex issues. I think there are some fundamental issues that destroy community. I'm not interested in simply a quantitative increase in housing or services—I'm interested in community building; that's my main agenda. That means taking on tough issues and having open discussions and mutual education around those issues.

I work pretty cooperatively with my students and with the community. Working on the field projects, my role is more like being a team leader than like being a teacher of a class. Certainly we have disagreements, but as a team and with the community we work them out. I have more experience than most of the students, but I need to use that experience to share that knowledge with people. It doesn't mean that I need everybody to agree with me specifically.

A number of years ago, when I was part of the Urban Analysis Group, we were meeting with community leaders from Roxbury [a well-organized, diverse, multiethnic neighborhood] and also labor leaders from a restaurant/hotel worker union in the city about some housing initiatives we were pursuing. A number of the leaders were either graduates of [our] program or current students. Our position on home ownership was very negative: "We don't really support that—we want social ownership of housing." The Roxbury people were saying, "Our community has a much lower rate of ownership than in any of the white communities, and we don't accept this. We want to be homeowners just like everybody else. We feel like having a lot of homeowners in our area will stabilize the community."

So we went around and around. Then we tried to figure out the advantages: "Stabilizing the community is a good advantage, something you want to do. Let's try to pull out the other things that people think home ownership provides that are not currently provided in the rental market." We went a few sessions on security of tenure, being able to personalize your environment, having a relatively predictable housing-cost outlay. What we came up with out of that process was a formulation for non-speculative home ownership, which has now been instituted through a number of the neighborhood housing development corporations in Roxbury. That, to me, is a concrete example of where I learned a great deal by not writing off the community's concerns just because they didn't fit into my political program. The entire group learned how to reformulate something so that it really worked for the community over the long haul.

We also didn't just say, "So you want home ownership? Okay, we'll just go work for home ownership." We said, "Home ownership is sometimes a way a lot of people get screwed in the speculative housing market." That went back and forth, and I think it was respectful on both sides, both sides learned, and we came up with something that we all felt really good about. (Forester, Pitt, and Welsh 1993:118–120).

Public deliberations may rarely produce outcomes that "everyone will feel really good about." I have tried to suggest instead, though, that we should not underestimate either the possibilities of public learning or the importance of the working through of past suffering, the ways that past loss shapes present political identity and its evolution. If we do not ask citizens to leave their pain at the door, and we reshape our deliberative processes in directions I have sketched, we might better enable both voice *and* mourning, mutual learning *and* public action, that recognize and respond to needs and interests more and miss and dismiss them less.

8

On the Ethics of Planning: What Profiles of Planners Can Teach Us About Practical Judgment and Moral Improvisation

Because the study of ethics is the study of better and worse practice, ethical inquiry should be practical—neither so idealized nor abstract that readers will find it "irrelevant."[1] To explore the complex ethical challenges that confront planners and policy analysts, we will examine excerpts from a profile of a young transportation planner discussing her work. We will see that not only are planners' ethical judgments inescapable in practice, but that they require a demanding and even principled form of moral improvisation as well.

Ethics Concerns Far More Than Decision Making

Ethical challenges in planning and policy analysis involve far more than "making the right decision." Ordinary language suggests as much. Rarely do we confuse "planners" and "decision makers," even when we speak of planners "informing the decisions" of elected officials. Because "*deciding to do* the right thing" is only one small part of actually "*doing* the right thing" (Rorty 1988), we should not narrow the scope of ethical analysis to decision theory (Storper and Sayer 1997, Howe 1994, Rorty 1988, Murdoch 1970). Furthermore, identifying "good reasons" for acting does not cover the broad range of ethical issues. As Amelie Rorty (1988:283) puts it, "Even if the reasons that prompt an action are appropriately justified, they usually underdescribe and underdetermine the detailed thoughtfulness required for appropriate action."

Planners do much work before decision makers act. We say, for example, that effective planning typically influences, informs, guides, shapes, or even focuses subsequent decision making. But if planning is not decision making, what else does the ethics of planning involve?

We could, for example, view planners' work as wholly separate from decision makers'. We might think of planners as advocates for particular communities, or as technical research staff who study issues long before they appear on decision-making agendas. But even if we reasonably distinguish planning analysts from decision makers, many planners nevertheless do work closely with decision makers and attempt to respond to "their needs" (both real and imagined).

Sources of Planners' Influence: The Ambiguity and Uncertainty of Decision Making

A better strategy of exploring the ethics of planning would have us examine how planners handle the inherent problems that decision makers face. Just to the extent that decision makers face inescapable and thus expectable problems—because they need help regularly assessing options, envisioning consequences, justifying choices, and so on—planners can both anticipate those needs and respond to them more or less well, ethically better or worse.

Where decision makers have questions, planners will have opportunities, but how will they anticipate and respond to them? The most important ethical questions here involve less the periodically incompetent or venal decision maker (although particular ethical problems arise then to be sure) than the structurally inevitable limits of decision makers' attention—limits that planners and policy analysts must always address. Those inevitable limits present decision makers with uncertainties and ambiguities no matter how "rational," skilled, or thoughtful they may be (Forester 1989, Krumholz and Forester 1990; cf. Howe 1994, Hendler 1995).

Two deep difficulties of decision making involve the quality of information and knowledge decision makers can have about (1) the streams of consequences, the "results," that today's decisions will generate, and (2) the values, interests, or preferences by which these "results" should be assessed, compared, discounted, and evaluated (March 1988). We can call these problems, following James March, the challenges of *uncertainty* about consequences and *ambiguity* about values, interests, and preferences.

These decision-making difficulties also echo a classical Aristotelian argument in ethics. Because value in the world is plural and incommensurable, because acting well depends on keen perception and responsiveness to such value in unique and particular circumstances, acting well depends primarily on the exercise of good judgment and not primarily on the skills of making correct calculations. Accordingly, to cultivate right action—to encourage anyone to act well—we need to cultivate practical judgment rather than simply to impart technical skills (Nussbaum 1990a).

Models of Right Action: From Calculation to Judgment

Consider several differences between judgment and calculation. Ethical calculation depends on prior givens, both data and techniques, and a certain fit between the two, as well as requiring proper responses to the problems of uncertainty and ambiguity. Ethical judgment, in contrast, works to fit action to circumstance, to see general principles in the light of contextual details (and vice versa).

Based on strong assumptions about relevant data and the rationality of actors, calculation provides crisp but potentially oversimplified solutions. Based on a consideration of general norms and unique detail, judgment provides nuanced but potentially indeterminate conclusions.[2]

Arguments for calculation prize rigor and consistency but sacrifice realism. Arguments for judgment may prize realism but sacrifice rigor and consistency.

Obviously both of these accounts of ethical practice are problematic. The appeal to calculation satisfies a rationalist intuition that there might be "one best way," one right way, to proceed in any messy and conflictual situation. The appeal to judgment satisfies an equally deep intuition that the moral world is not neat, that fulfillment and loss do not cancel each other out, that values are plural, diverse, and incommensurable, not simply superficial versions of some underlying and unitary value expressible in terms, for example, of money or pleasure (Nussbaum 1986).[3]

The appeal to calculation promises moral precision, but quickly runs into trouble when confronted with the messiness of everyday action. The appeal to judgment promises a perceptive and responsive appreciation of daily complexity, but it gives up the ideal of a formal moral calculus, arguing instead that such a formal calculus is a deeply mistaken ideal, an ideal not just unattainable but undesirable and misleading too (as if parents could make daily decisions about their children on such a basis). As Martha Nussbaum (1990a:70–71) puts it, "Good deliberation . . . accommodates itself to the shape that it finds, responsively and with respect for complexity. . . . A doctor whose only resource, confronted with a new configuration of symptoms, was to turn to the textbook would be a poor doctor; a pilot who steered his ship by rule in a storm of unanticipated direction or intensity would be incompetent."

So we have no ethical cookbook, and appeals to judgment are hardly more reassuring than those to calculation, particularly for those who hope to discover and "apply" some best ethical method, procedure, or technique. David Wiggins (1978:150) suggests we face the music: "I entertain the unfriendly suspicion that those who feel they *must* seek more than [the Aristotelian account of practical reason] provides want a scientific theory of rationality not so much from a passion for science, even where there can be no science, but because they hope and desire, by some conceptual alchemy, to turn such a theory into a regulative or normative discipline, or into a system of rules by which to spare themselves some of the agony of thinking and all the torment of feeling and understanding that is actually involved in reasoned deliberation."

Practical Judgment and the Character of Moral Improvisation

The human calculator tries to be a skilled engineer, to find a best solution, "to solve." In contrast, the person of practical judgment tries to be a sensitive and principled moral improviser: to attend to both the unique details and the general norms and principles relevant to *this* complex circumstance.

What can it mean to improvise where moral issues are concerned? Nussbaum (1990a:95) extends the metaphor of theatrical improvisation as an "image for the activity of practical wisdom" this way:

> Does [the moral actor], in learning to improvise, adopt a way of choosing in which there are no principles and everything is ad-hoc? (Perhaps: in which every-

thing is permitted?) The image of the actress suggests how inaccurate such an inference would be. The salient difference between acting from a script and improvising is that one has to be not less but far *more* keenly attentive to what is given by the other actors and by the situation. You cannot get away with doing anything by rote; you must be actively aware and responsive at every moment, ready for surprises, so as not to let the others down.

Today the growing literature on so-called rational choice informs questions of moral calculations, but we seem to know less about moral improvisation. Considering practitioners' accounts of their work, we hear few accounts of the unproblematic use of technique, many stories of contingency and surprise—of conflicting norms and actors and the need for response, many stories of discretion in the face of ambiguity, many stories of the fluid coming together of provisional agreements, timely opportunity, limited resources, and pragmatic responses. We need to learn more about how planners morally improvise in real, nonideal settings.

Perceptive moral improvisation, Nussbaum argues, requires attention both to principle and to detail, both to encompassing norms and commitments and to specific circumstantial particulars. As she (1990a:95) writes, "The perceiver who improvises morally is doubly responsible: responsible to the history of commitment and to the ongoing structures that go to constitute her context; and especially responsible to these, in that her commitments are forged freshly on each occasion, in an active and intelligent confrontation between her own history and the requirements of the occasion."

We need to explore not only how planners respond to standing norms and principles, but also how they attend to situational details that matter. Far from being "free to do whatever they want," good moral improvisers must respond to both overlapping goals and mandates, norms and obligations, *and also* to the uniquely significant particulars and details that make each case what it distinctively is.

Once we view planning as more than simply constrained optimization, we can listen to planners in a new way. The thick and murky character of practitioners' stories becomes now not so much mud to slog through, but for astute listeners a fertile source of practical insight instead (Abrams 1991; O'Neill 1989; cf. chapters 1, 2, 3, 4 above). Indeed, these practice stories show us the ups and downs of moral improvisation: the difficulties and contingencies of practical judgment in a fluid and interdependent political world.

Practice Stories Illuminate the Work of Moral Improvisation

Let us consider how these stories can teach us about ethics in planning and public policy settings. Listen first to Kate, a county transportation planner:

The work I do has two components. The first is that I'm an administrator for a federally funded metropolitan planning program. . . . I do a lot of the actual administrative tasks of budget preparation, program development, monitoring expenditures, remitting payment requests—the general things that need to be done to keep the program going and the dollars flowing.

The other part of my job is technical. I actually head up or supervise most of the technical work that gets done for transportation planning for the county. In recent years, that's included the preparation of a highway plan for the county; the preparation of two or three subarea studies, actual transportation and land use studies of specific highway corridors in the county that are undergoing rapid development or are tremendously congested or threatened in some way; and also data collection activities needed to support the planning that we do. I've written or done a lot of these studies myself.

Sometimes it's frustrating to coordinate the two tasks. The administrative work is work that has to be done, in a sense, to keep the program going, to meet the requirements that are set by the state and the federal government. That's sometimes frustrating because it takes time away from the objective of the program, which is to do long-range transportation planning. It's not a major conflict, but there are times when I'm revising the work program for the fifteenth time, and I just wonder how useful that is when what I really should be doing is spending some time analyzing the data we've collected or working on one of the long-range plans in the work program."

Kate echoes a familiar problem of professional practice: trying to do the transportation plans, the subarea studies, the basic data collection that needs to be done *and* to "keep the program going and the dollars flowing." Clearly there is no cookbook answer to how best to "coordinate these two tasks," and the difficulty of doing it, sometimes, is "frustrating."

Conflicting Obligations and Values to Be Honored

"I just wonder," Kate says, "how useful [revising the work program for the fifteenth time] is when what I really should be doing is . . . analyzing the data . . . or working on one of the long-range plans." If we listen closely, though, we can see that Kate details an almost bewildering number of conflicting obligations:

- The broader institutional "requirements set by the state and federal government"
- The very specific details of "remitting payment requests"
- The more intermediate obligations of so-called budget preparation and program development, perhaps vague in name because they are intrinsically creative in character
- The technical requirements, which are as obligatory as any of the legal requirements, so Kate tells us about the "data collection activities needed to support the planning that we do"
- The organizational obligations of supervision, so Kate says, "I actually head up or supervise most of the technical work that gets done for transportation planning in the county"
- The specific obligations that flow from the current "work program"
- What Kate tells us is "the objective of the program" itself: to do something called "long-range transportation planning," an objective whose interpretation will shape just how the "work program" will actually work out

Even if we recognize these many kinds of obligations that Kate faces every day, we might still imagine that no "improvisation" is necessary, if, for example, these obligations might somehow nest neatly and logically under one another. Federal requirements, for example, might shape state requirements, which shape county requirements, which specify budget requirements and program objectives, which in turn shape the work program, which in turn shapes data collection requirements next Tuesday. In this ideal-typical view, obligations and duties—the planners' "what I ought to do today and tomorrows"—fit together in a neat, top-down hierarchy, leaving little room for improvisation or ethical judgment.

The deep appeal of this hierarchical picture can be difficult to shake. Maybe the world does not really operate this neatly, the argument might go, but it should! Furthermore, this orderly view can reflect not only the perspective of outsiders—who might be blind to the internal messiness of actual practice—but insiders as well, who think of such an ordering of responsibilities as an ideal, if unfortunately not as a crisp, description of reality.

But this ethically tidy hierarchical picture of practice belies reality. Virtually no one talks this way about the duties of daily work. Virtually no research on planning or policy analysis suggests that this picture does justice to the real world, for very well-established reasons. Hardly isolated from other institutions, planners work closely with them; the political environment of practice changes as elections and coalitions come and go; relations with advocacy organizations, citizens, and other agency staff are fluid, and they shift as political administrations transform, as international trade supports or weakens regional economies, as agency leadership changes, as priorities shift. Interests and goals are multiple, ambiguous, and conflicting, and so on. The ideal-typical hierarchy of mandates becomes a tangled web of obligations in real practice.

Ambiguity and Rule Richness Produce Deliberative Space—and Make Ethics Inescapable

Notice now that as the rules that govern practice become fewer in number, less in conflict with one another, and more definite, the less discretion practitioners will have; but the more diverse and ambiguous—and thus open to multiple interpretations—the rules, the more discretion practitioners will have. This seems obvious, because the more that the rules control, the less discretion remains, and so the less room exists for ethical deliberation and judgment. But this seems counterintuitive too, because as practitioners face multiple and conflicting rules, more complexity, diverse goals and interests and values to be honored, they may have *more* discretion institutionally—because of such complexity—more room and even greater necessity for ethical judgment and deliberation.

Now we can better appreciate the complex challenges that Kate, the transportation planner, faces. She has no cookbook to tell her which rules to follow in what sequence and at what times. She must try to do them all justice—or fail: fail to do competent data collection or analysis, or fail to meet the state and federal procedural requirements, or fail to administer her ongoing programs effectively.

Managing Value Complexity

So how can planners and policy analysts do this complex work of judgment and deliberation that their rule-rich environments require? If we listen carefully, we can appreciate how Kate responds to the many obligations and value complexity of her tasks:

> The primary long-range plan I've been working on, which has just been finished and has gone to the printers, is the county highway plan. The plan takes a comprehensive look at the major roads and highways in the county through the year 2010. It looks at a lot of information about the demographics of the area, the historic traffic patterns mostly on state and county roads within the county boundaries, and does some projections as to which roads may be experiencing capacity problems between now and 2010. It's basically a twenty-year planning audit, recommending both physical improvements for the roads—for example, which roads might need to be widened—and also trying to take a more comprehensive look at other ways to meet the transportation needs in the county. It considers ways of reducing demand or promoting transit or car pooling for work trips, rather than just constantly widening the roads, which has proven to be somewhat futile.

This story, simple enough on its surface, can begin to teach us about the ethical strategies Kate's planning staff use to honor multiple values, to respond to multiple obligations, and to reconcile potentially conflicting goals.

First, Kate notes an *administrative strategy of dividing attention:* assigning staff to work on several projects at once. The plan just finished and sent to the printer's was the primary long-range plan on which she had been working.

Second, Kate notes their efforts to *attend to diverse problems and values*. This plan takes a "comprehensive" look, she says.

Third, though, because that comprehensive look must be selectively attentive, it focuses on a *functional ranking* or classification system of "major roads and highways," and it does so within a selected time frame, through the year 2010.

Fourth, Kate indicates that the long-range plan *organizes disciplined (and disciplinary) attention*—attention both enabled and justified—by analytical considerations of the "demographics" of the area, historic traffic "patterns," and some "projections" of possible problems.

Fifth, we learn that the plan *responds to the limits of yesterday's investments*. Acknowledging how yesterday's commitments of resources will lead to problems tomorrow, the plan anticipates the "capacity problems" of the existing road system.

Sixth, Kate notes that the plan itself is hardly idiosyncratic or blindly improvised; it borrows from *standard procedure,* a conventional accounting ritual: "basically a twenty-year planning audit," she says.

Seventh, Kate teaches us that the plan does not respond simply to a single all-embracing value of improving transportation in the county. It *responds more specifically to distinct, and so potentially conflicting, goals:* it recommends both "physical improvements" as well as "other ways to meet the transportation needs": "reducing demand," "promoting transit," and "car pooling," for example.

Eighth, Kate notes that her staff tries to *learn from the successes and failures of the past* as well. "Just constantly widening the roads . . . has proven to be somewhat futile," she tells us.

Effectiveness and Ethical Evaluation: Of What?

Of course, we do not know how well Kate and her staff do this work. But asking the question of effectiveness too soon risks missing a crucial point. We cannot possibly begin to evaluate Kate's practice until we recognize the diverse challenges that she, and planning analysts like her, are called on to meet in their everyday work.

We cannot evaluate Kate's practice, for example, in terms of her analyses of physical improvements alone, because an important part of her work involves analyses of nonphysical strategies. We cannot evaluate her practice in terms of contributions to just one plan, because we know that her job demands work on several other projects too. We cannot evaluate Kate's practice in terms of disciplinary competence alone (she may be a superb technician), because we know that a plan she is drafting must be comprehensive too (is she too narrowly focused?). So our problem, here, in evaluating what Kate does—in assessing the ethics of Kate's practice—is not that she may do too little, but that she may do too much!

Improvisation and Learning About Value: Deliberation

Kate must somehow act on a diverse, plural, and ill-defined set of concerns and values—values that may be, in any actual case, honored or ignored: physical and nonphysical needs, technical and nontechnical issues, county transportation needs and problems of agency administration as well. Kate must pay attention to plural and not easily commensurable values, and she can hardly expect to "know what to do" by using some cookbook technique. She must learn about the values at stake in her work just as she learns about the county too.[4]

So we can hardly begin to understand planners' and policy analysts' work as simply assessing, selecting, or recommending "means" to well-defined and politically given ends. Instead, their practice is inescapably deliberative, necessarily exploring and probing goals and ends as well as means.

Kate, for example, must explore the character of the ends themselves because they can have specific meanings in these cases: administrative efficiency, systemic versus narrow analysis, timeliness, technical rigor, physical upgrading, policy innovation, and so on. Not only must she try to reach or achieve goals, but more significantly and subtly, she must also clarify and specify—give actual content to—those same goals as well. She must do the work of "norm setting" as well as norm following (Majone 1989, Vickers 1995, Richardson 1990). If we miss how planners and policy analysts set norms and shape goals, then we will miss important aspects of their practice.

Improvisation and Perception of Particulars: Assessing the Facts That Matter

But Kate must do even more than specify ends. Even with ongoing attention to "goals" and "needs" and state and federal requirements, Kate must study the unique particulars of her county. She tells us:

> Unusual solutions will be needed to solve new problems. For example, we're on the eastern shore of a river, and there are two bridges that connect our county with the counties on the other side of the river. One of the bridges is experiencing some delay on it during the peak hour, and the Bridge Authority is making noises as if they want to build another bridge at God-knows-how-many-million-dollars cost.
>
> Really the problems on the bridge are confined to two very specific time periods: the morning peak hour and the afternoon peak hour. A big part of the congestion problem is that there are a lot of people who work for a major corporation here in the city who live across the river. So actually you have one employer who employs somewhere between 20 or 30 percent of the people who cross the bridge at peak period. That presents an interesting situation for trying to get an employer involved in things like van pooling or providing incentives for their employees to car-pool—like closer parking spaces or some other incentives.

Kate's story about "the congestion problem" on the bridge is deceptively simple. She moves quickly from telling us that one of the bridges is "experiencing some delay" to telling us about the looming institutional power of the Bridge Authority, "making noises as if they want to build another bridge at God-knows-how-many-million-dollars cost."

But "really the problems on the bridge," she goes on to say, "are confined to two very specific time periods," the morning and afternoon peak hours. Is the problem, then, we might ask, political power (the threat of the Bridge Authority's expanding its domain: a large hammer in search of a nail?) or the timing of traffic flows and delays?

Quickly we learn that "a big part of the congestion problem" is not just timing, but the particular commuting patterns of "a lot of people" who live across the river and who happen to "work for a major corporation here in the city." Estimating that these employees alone account for 20 to 30 percent of peak hour loads, Kate raises possibilities of "trying to get the employer involved," trying to persuade, interest, or influence the employer to consider programs like "van pooling" or "providing incentives for their employees to car-pool."

So "the problem" of the bridge can really be a problem of gaining a major employer's cooperation—to van-pool or stagger employee hours—or really a wholly different problem of institutional power and turf. "The facts of the case" simply do not speak for themselves.

The facts are no more "given" than the goals are, which means that Kate and her colleagues have their work cut out for them. Faced with an apparent problem of congestion on one of the bridges crossing the river, they must not just "get the facts," but they must get the facts that matter. They need to get not just information, but relevant information, and to do that, they must exercise good judgment and not simply follow any predetermined data collection method (Kronman 1986–1987).

The Priority of Practical Judgment

As the truism has it, before planners can solve a problem, they need to construct or formulate just what "the problem" shall be taken, for practical purposes, to be: they need to pay careful attention to the unique particulars of the case. In Kate's context, these unique particulars might range from the personalities of labor-management leadership in the corporation to that corporation's production and scheduling technologies to the political structure of the Bridge Authority to the physical configurations of the bridge's feeder and access-exit systems.

Even a few paragraphs from Kate's account reveal the practical-ethical complexity of her work. Kate cannot simply calculate solutions to problems, because she must pose those problems clearly before she can solve them. She must judge which facts are relevant to any problem she faces, for the relevant facts do not walk up and introduce themselves.[5] Kate must also simultaneously assess those facts in the light of conflicting and ambiguous goals: improve transportation, be efficient, be responsive to the public, serve the poor, and so on.

So Kate has two prior problems of her own. She must be, in Nussbaum's (1990a) sensitive phrasing, both "finely aware" and "richly responsible." Paying careful attention to detail, looking carefully, listening carefully, she must be finely aware of the unique particulars of this situation, this locale, this negotiation, this political circumstance. Paying careful attention as well to encompassing principles and mandates, she must be richly responsible to the multiple, conflicting, and inevitably ambiguous responsibilities, goals, and obligations she inherits and must work to honor in this case.

The very complexity and messiness of Kate's account can teach us about the creative moral improvisation she must do at work. We can call this work "practical judgment" and "deliberation"—to mark its noncalculative or precalculative character—but the labels themselves can obscure as much as they clarify (Majone 1989, Reich 1988, Wiggins 1978). Similarly, we can think such work is "common sense," but as a planner in San Francisco put it wonderfully once as he described conflictual planning settings, "Common sense isn't always that common!"

Kate and planners like her need not only to improvise strategically as they face complexity, but they must improvise morally too.[6] They must learn about value in the world—what is at stake in this action or that inaction—just as they must learn about fact. If they succeed in their "subarea studies," for example, they will learn about threats and dangers and opportunities, and they will see issues—value at stake—more clearly than they have. If they succeed, they will have come to see more than they saw before about what is really important in each "subarea." Especially when the apparent goals and destinations—a mandate to "do transportation planning," for example—are vague and ambiguous, planning studies will often address "where to go" as well as "how to get there" (Vickers 1995).

Planners can, need to, and do learn about value in the world. We can now understand how they can do that, for better or worse, if we examine carefully their deliberative practices of listening and encouraging participation.

Listening and Learning About Value

In practice settings, listening carefully enables not just some vague "better communication," but a far more important recognition of what others really value, beyond their wants of the moment (Forester 1999b). Listening well, planners can probe beyond another's literal words—so they can pay attention instead to the speaker's real concerns, always evaluated in the institutional context at hand.

The power of recognition that comes from a listener's, "Oh! So what you really mean is . . ." grows not simply from their clarification of meaning but from their fresh appreciation of value, of what matters, seen now in a new way. Planners must listen to learn *not just about* a resident's or a developer's "values," but about their hopes and fears; not just about their demands and "positions," but about their underlying interests; not just about their expectations but about their entanglements, interdependencies, and vulnerabilities; not just about their expressed "reasons," but about their needs for self-esteem, recognition, perhaps at times "saving face" too (Forester 1989:chap. 7, Storper and Sayer 1997).[7] Listening well in practice protects planners from what Robert Coles (1988) has called "the rush to interpretation"—a rush made more tempting when their time is limited, when they distrust others' motives and stories, when they see others as threatening or adversarial (cf. chapters 1, 2 above, Kochman 1981).

Participation and Deliberation

Participatory planning processes can promote deliberation, not just mirror public wishes. From mediated negotiations through action research to representative working groups, such processes can enable participants not only to bargain and trade resources, but to learn and grow as citizens too. Participants can not just claim resources and "get more for themselves," but they can learn about the Other—about their common destiny, about false suspicions and false promises, and about their new options for strengthened relationships and common action.

We see the deliberative character of participatory processes most clearly in mediated negotiations in planning. Working as skillful mediators, the planners' job is not to use that magic wand of public polarization, the formal public hearing, where all loud voices might be heard but where no dialogue or collaboration will be possible. Instead, the planner-mediator's job is to ensure that affected voices not only speak but actively craft mutually acceptable agreements together, avoiding exclusionary deal making as they go (Susskind and Cruickshank 1987, Susskind, McKearnan, and Thomas-Larmer 1999). Such mediators can encourage conversation and negotiation in which parties typically come to see issues in new ways, come to see other parties in new ways, come to see their own options and thus themselves in news way too (Ferraro 1992, Forester 1999b). In doing so, participatory planning processes promise not just loud posturing but learning and, at times, the transformation of disputing "I"'s into a tentative and fragile polity of a "we" going on together despite our differences (cf. chapter 5 above, Barber 1984).

Important questions remain: Are deliberative mediated-negotiation processes, for example, prone to manipulation and co-optation? How might deliberative processes vary sociologically? How can deliberative practices encourage or discourage varying forms of recognition of others? How can organizational form and ritual structure shape political participation, and thus deliberation and learning about value?

Arguably, planning practice leads planning theory here, for in practice such learning about value happens all the time, if for better and worse in different circumstances. But just this needs study: Why do skilled negotiators routinely talk about the informal aspects of getting to know their counterpart negotiators "away from the table"—and what does that suggest for the ritual structuring of deliberative and participatory processes in planning? How can organizational forms of storytelling have such powerful effects on community members' senses of membership, visibility and recognition, memory and re-membering—and what can that teach us about the staging of deliberation in planning (Myerhoff 1988; chapter 6 above)?

If deliberation is an inescapable, deeply necessary but contingently structured part of the practice of planning and policy analysis, what forms can it take? How can planning analysts encourage productive public deliberations, strengthen rather than weaken them, and protect them from the play of dominating power?

The Politics of Moral Improvisation

One might object that our portrait of planners as moral improvisers is too silent about politics. Won't moral improvisers always be vulnerable in the face of power unless they are armed with analyses of agenda setting, of the creation of desire and of ideology, to guide them and the skills to respond? Put slightly differently, can moral improvisation and practical judgment be "critical" and pragmatic too?

We can turn this question around: Do we have any reason to think that radical democratic planners and community organizers, for example, do not also, and necessarily, have to be moral improvisers in their actual practice? Mustn't they necessarily attend to both the particulars of the communities in which they work, including their local traditions and rituals, as well as to the radical democratic traditions inspiring their senses of social and political possibility?[8] In a nutshell, then: academics can theorize, but practitioners must improvise. But how *should* they do that in planning and policy analysis settings?

To be more than an opportunistic deal maker, for example, the morally improvising planner should counteract exclusionary power by working to include representative voices of affected stakeholders; by anticipating self-protective behavior and claims from powerful actors; by exposing the systematic suppression of data; by resisting rationales of resignation and invoking potentially radical, if also traditional, values of agency, respect, dignity, and so on. The morally improvising planner who ignores the suppression of citizens' voice or data weakening the claims of the powerful would be willfully blind, hardly responsible to the basic obligations of public-serving planning.

If planners should be "doubly responsible" to general principles and unique particulars, they need to be attentive to two corruptions too—those of mandates and of facts. They must anticipate inherited exclusions structured by past injustices. They must anticipate citizens' or developers' or politicians' suppression of unflattering data. The morally improvising planner cannot decide unilaterally on solutions before assessing the problems at hand, but neither, as we shall see, can this planner be one for whom "anything goes!"

Now, do most planners today improvise practically in these ways, challenging exclusionary power and the discouragement of stakeholder participation? Probably not. *Should* most planners worry about exclusion, the suppression of data, the control of development to line the pockets of the few? Of course. Our purpose here is not to describe the lowest common denominator, but to explore what such improvised practical judgment involves at its best.

Done at its best, moral improvisation in planning will be neither exclusive opportunism, with its willful blindness to the legitimate claims of others, nor simple rule following, as if planners could use "gimmicks for justice" no matter what the situation at hand involved. At its best, moral improvisation will be perceptive, receptive, and dialogic—recognizing the value of the particulars in this case, recognizing the claims of conflicting parties, recognizing the traditional claims of the past no less than the anticipated claims of future generations. *Such improvisation and practical judgment presumes the need to learn about value* before any "solutions" can be discussed; it requires an ongoing search for value—for the needs, dangers, and opportunities there to be recognized, there to be acknowledged despite the play of power, bluffing, and manipulation—for the value there to be acknowledged in a world of injustice.

Acknowledging conflicting claims of citizens and rich and incommensurable sources of value in the world, the morally improvising planner at her best does not act unilaterally.[9] She is a participant in the moral world she perceives, a participant who must be sensitive to the political structuring of others' claims of value. Listening and learning, she must interpret and reconstruct problems, probe strategies that "do justice" to what she sees, and encourage deliberative working agreements that enable citizens to act together.

Does Moral Improvisation Mean "Anything Goes!"?

Does this mean that anything goes—that our perceptive, engaged, listening, and probing morally improvising planners—women and men of practical judgment—can do anything they please? Hardly. But we have formulated no rules, put forward no clear prescriptions: so why isn't this account of practical judgment just a recipe for chaos, arbitrariness, and resignation in the face of power?

Consider an analogy here: what we might do to seek advice when we have an important decision to make. Let us say we are concerned, confused, maybe even troubled by an opportunity that has come our way—to take a new job, for example. We might turn to a friend less to get expert advice or automatic agreement, more to get a better sense of what this "opportunity" could mean for us and those we care about. We turn to a friend because we think he or she will listen closely, remind us of important things we may be forgetting, see an issue clearly where we are confused, and so on. With a difficult practical judgment to make, we might well hope that a careful conversation with a friend would help us to deliberate well, to learn about the issues that will really matter here, to consider what speaks for one course of action or another. Concerned about real action and attentive to its complexity, its ambiguities and uncertainties, we might turn to a friend for an ear, a heart, and a mind. We might do this imaginatively, of course, as we think, "She would say . . ." "He would wonder about . . ." as we consider still other issues.

In such cases, can "anything go" for the friends to whom we turn? Because they must improvise their responses to our unique case, can they do anything they please? Clearly they cannot, because they need to listen to our particular concerns, to attend to relevant issues they may see but that we may have missed, to ask about consequences we have poorly considered, to remind us of commitments we have made, and to help us see clearly what we will be doing if we alter those commitments. Our friends are not free to "do anything they like," for they are doubly responsible to recognize and respond to the actual richness of the "problems" we bring to them—a richness involving the details no less than encompassing commitments and principles too.

A similar case holds for practical judgment and moral improvisation in the public domain. To say that one rule does not govern in all circumstances, that no one technique can always be used, is not to say that "anything goes." To say that practicing planners and policy analysts must be moral improvisers does *not* mean they can do whatever they please. It does not mean that they can be blind to the claims of the powerless, that they can ignore the needs of those most in need. It does mean that they must be women and men of judgment; they must be perceptive, probing, engaged, attentive to the blinders of power and the suppression of voice; they must have enough moral imagination to be able to listen, to recognize and respect difference; they must have enough political imagination to be able to foster real deliberations, real considerations of common, interdependent destinies, identities, and possibilities.

Developing Judgment

To pose problems well before they craft responses to them, planners need to be, in Nussbaum's terms, as "finely aware" of detail as they are "richly responsible" to diverse mandates, goals, and obligations. Such context-sensitive, improvised practical judgment involves much more than applying classroom tools and techniques. Listen once more as Kate reflects on how she has changed and how she developed a more imaginative, creative capacity for practical judgment:

> I have a greater ability, I think, to look at the whole picture. I have a better sense when I start a project of where it's actually headed and what's going to happen. When I first came here, I was given pieces of things to do, which is how a lot of people start out; part of a zoning ordinance or part of a master plan. It took me a while to think more comprehensively about how all those pieces were going to fit together, and to know more when I start out as to where I'm headed. I think about what kind of report I'm actually developing and who the audience is, rather than going through saying, "This is the standard way we do projects like this."
>
> I've changed some projects midway, where we started doing the standard list of things—background chapters or whatever—and it just didn't fit the paticular project. In these cases, I throw some of it out and reorganize it in a way that puts the emphasis more where it belongs.
>
> The highway plan is radically different from the way it looked last May. In the interim, I was given responsibility for revising it, and I said, okay, if I have that responsibility, it's going to be revised in a way that makes sense to me, and hopefully to other people. I can do this as a result of having more confidence and being able to look at the bigger picture."

Kate has learned, she tells us, to "think more comprehensively," to imagine "how all those pieces were going to fit together." She has learned about the fit between a general direction, "where I'm headed," her particular contribution, "the kind of report I'm actually developing," and "who the audience is." But more too: she has learned, and she helps us to learn, that "the standard way we do projects" will not always do: "it just didn't fit the particular project."

Going on where the standard way would not do, Kate had to improvise, to "reorganize it in a way that puts the emphasis more where it belongs." So Kate takes responsibility in a broader conversation where her acts need to make sense not only to her, but "hopefully to other people" too. Finally, Kate suggests when practical judgment and deliberative practice can flourish: she could revise the highway plan to be "radically different," she tells us, because she developed a greater trust in planning with others, "having more confidence," she says, and developing a finer, more imaginative and responsive view of projects, "being able to look at the bigger picture."

Conclusion

We would need more information to determine if Kate had acted well or poorly on particular projects. But we have not been evaluating her performance in a specific case. We have instead explored how we can learn about the ethics of planning practice from planners' accounts of their own work.

Although practitioners' stories do not provide rules and recipes that detail in five easy—or fifteen difficult—steps, "How to Plan," those stories can teach us about the messiness of everyday work: the multiple and conflicting responsibilities that planners have and the complexities of the problems they face. Those accounts reveal more than messiness, too, for they show us the challenges and opportunities of practical judgment—in daily deed, the "doubly responsible" challenges of moral improvisation—that planners face at work.

We need a much more inclusive view of ethics than one reduced to decision making or acting on so-called dilemmas. We need to recognize that value in planning settings comes in plural forms—some quantifiable, some obviously not. Although economic analyses often assume the commensurability of values, that assumption can be unwarranted, or worse:

a recipe for producing resentment (e.g., when a citizen's deeply felt values are treated as purely financial matters). If planners should often respect value in the world as plural and incommensurable, those planners will need to learn how to do that, as they recognize that no decision-making calculus will allow diverse values to be easily added and subtracted to yield net gains or losses.

Where plural and incommensurable values are at stake, planners' judgments and plans will be selective and involve loss, whether planners like it or not (Davy 1997). If planners are blind to the roads not taken, they can make their audiences more suspicious, more stupid, more vulnerable than they might otherwise be. To avoid fueling resentment and contempt, damaging relationships further, planners and policy analysts must recognize and learn to articulate necessary and tragic loss instead of denying or hiding such loss behind so-called net benefits (Nussbaum 1986).

The moral complexity of planners' accounts, then, can teach us about the ethical character and ethical challenges of planning practice. We learn how practical judgment must precede, if not altogether preempt, calculation. We learn how problem formulation and construction drive problem solution. We learn that when mandates are ambiguous, as they often are, when the significance of the facts is not immediately obvious, deliberation must precede decision making. Like all other practical actors, planners must *learn about value as well as about fact* as they explore what is at stake in any specific case—why any projected consequences, for example, will be consequential.

We learn, too, that when time and information are in short supply, when politics and conflict are ever present, and when yesterday's solutions are not quite right for today's problems, planning analysts will have to improvise care-fully and creatively rather than simply collect "the facts" or "follow the rules" mechanically. A value-free planning practice—one blind to bias, dignity, voice, inefficiency, ineptness, illegitimacy, and so on—would be valueless, or worse: simply resigning itself to enabling "might" to make "right" (Fischer and Forester 1987). Which facts to take as significant, and which rules and responsibilities, goals and obligations, promises and understandings, to fulfill in what ways—these are inescapably moral matters that practitioners must face all the time. In this deliberative work we find the ethical character of daily planning practice, the true ethics of planning.

Afterword

By interweaving stories with commentaries, this book has explored practice and theory together. We began with quite ordinary work—an excerpt from a planning staff meeting and the account of a young planner discovering the politics of land use development—but we moved on to consider quite extraordinary work too. We explored the accounts of housing and community development planners who worked uphill in the face of racism and who found opportunities for successful collaboration in the oldest of bureaucratic battles between city agencies. We next explored the stories of several planners doing similarly deliberative, complex, and ambitious work: these practitioners built a design and development consensus that joined architects and archaeologists, economic development and tourism officials. These planners reached far "beyond dialogue" to make participatory action research work; they worked courageously, too, in the shadow of historical inequality, trauma, and past loss.

These practitioners' accounts of their deliberative practice are hardly reducible to one-liners or sound-bites—simple labels of "progressive" or "controlling," "good" or "bad" practice. Just as planners who are stuck or puzzled need fresh insights or reminders to go on, so do students of planning need practical stories with moral depth to provide substance for their inquiries. Practitioners without insight will be callous, barely competent, if not altogether ineffective; students and theorists of planning without the moral perception—the appreciation—of what is pressing in real cases will be naive and irrelevant, if not unwittingly condescending and disrespectful too. The friction—the difficulties and challenges—of excellent practice can lead to fresh lines of theoretical inquiry, and insightful theorizing, in turn, can provide suggestive avenues for practice (Forester 1999a,b).

These practitioners' accounts are not case histories, but they are windows onto the world of planning possibilities. These practice stories provide insights rather than simple answers, reminders rather than instructions, examples rather than hypotheses. They teach in part by showing complexity and detail, political entanglement and moral perception, not by providing deductive arguments, although the accompanying commentaries may do that.

We have seen *through* a wide range of cases—in which typically future consequences of public actions are uncertain and the responsibilities of public officials and professionals are ambiguous—that planners and policy analysts can foster public deliberation and learning in three complementary but distinct ways.

First, planning analysts can encourage technical inquiry about available strategies and employ diverse analytical methods of project and policy analysis. Bringing professional expertise to bear, planners may present economic analyses of market viability, demographic projections of job or population growth, or architectural or urban design renderings of complex project proposals. Proposed sensitively with an eye to timing and political context, technical arguments can powerfully contribute to deliberative policy and design conversations rather than depoliticize public issues (Majone 1989).

Second, planning analysts can encourage explicit value inquiry: the careful analysis of costs and benefits, obligations and responsibilities, charters and mandates, goals and values to be honored or respected, protected or defended in a particular planning process. Such learning about value often takes the form of assessing how the consequences of alternative actions can be consequential: how, that is, they might matter, how and why they can be important. Even if only a few people will be affected in the short run, will longer-term consequences be substantial (after the election? after the next "fifteen-year flood," after the chemical moves through the food chain?)? These chapters have shown us that one fascinating part of planners' and public policy analysts' work involves their evaluative, moral inquiry about the future of our communities. We have seen that insightful practitioners can do work that our theories of planning, public administration, and political science barely do justice to. These practitioners must learn about value on the job, not just about consequences but about consequentiality, about threatened public value or human dignity, about opportunities to create beauty or autonomy. These practitioners' stories should prompt us to give still more instructive, theoretically insightful accounts of the moral inquiry that planners and policy analysts must do all the time in practice.

Third, planners and policy analysts foster deliberation about affected citizens' worries and fears, hopes and loyalties, commitments and self-images. Such "learning about social identities" is particularly important when planners face issues of race or gender, culture or ethnicity, for example, whether in neighborhood planning or international "development" contexts. Without skillful interpretation in sensitive listening to others' stories, we have seen, planning analysts can hardly build the working relationships with others they will need to get real work done together. This learning about others can take many forms: perhaps Baruch Hirschberg's "diplomatic recognition" in adversarial environmental and transportation planning negotiations, or the still more complex working through suggested by Mary Jo Dudley's work in the face of past trauma. Deliberative practice and participatory processes will fail if planning analysts pay so much attention to technique or "substance" that they ignore and dismiss the history and culture, the self-perceptions and deeply defining experiences, of the citizens involved.

These three faces of public learning present distinct and predictable challenges with very practical implications. Planning analysts who cannot carefully distinguish these three forms of deliberative learning will risk failing in their work altogether. They may be technically skilled but insensitive, and then dismayed when another's anger overwhelms their otherwise technically competent work. They may be sensitive networkers, but if they are incompetent technically, few people may seek their advice. They may be astute observers of community need and political mandates, but if they cannot build trusted relationships with others, they will end up talking to themselves, with little impact on public policy and public welfare. As I have argued throughout these chapters, effective planning and public deliberation require action-oriented learning about instrumental strategies, political ends, *and* social identities as well. These critical and pragmatic inquiries may be required simultaneously in complex settings. Through cases in the United States and abroad, I have shown how planning analysts have done this work, for better or worse in different cases, in quite ordinary and extraordinary ways too, so that we could imagine still more insightful and effective future deliberative practice.

To be sure, some planners and analysts have used the pretense of public deliberation to manipulate or control affected publics (Flyvbjerg 1996, Yiftachel 1995; cf. Forester, forthcoming). Some use the appearance of public deliberation as a whitewash to hide secretive decisions already made (Arnstein 1969). Some do it and raise public expectations that cannot be satisfied (Baum 1980). In *Planning in the Face of Power* (1989), I explored the ways planners could manipulate citizens by providing them with official, even professional "misinformation" (cf. Flyvbjerg 1998). This book has addressed a different problem, however: understanding how planners might actually and practically encourage participatory deliberative processes more or less excellently in real cases.

As these chapters have argued, public deliberation is always an iffy political accomplishment, a matter of political practice and performance that can never be guaranteed, that can never be simply a matter of good intentions. These chapters have explored many of the requirements of effective, educative, even transformative deliberative processes: requirements involving issues of learning and recognition, respect and listening, ritual and processual design, power and representation, historical trauma and past loss. Public deliberation in participatory planning processes is a contingent, fragile, vulnerable possibility in a precarious democratic society; it depends not on some virtuous "good planner," but on struggle and hard work, insight and imagination, moral sensitivity and political perception too.

Just as planners and policy analysts face the difficulties of public learning in these three ways, so do students of planning and politics as well. Provocative studies by Seyla Benhabib (1992), Iris Marion Young (1990), Jane Mansbridge (1992), Amy Gutmann and Dennis Thompson (1996), John Dryzek (1990), and Daniel Yankelovich (1991) urge us all to examine carefully the character of deliberative democratic politics. The work of these philosophers and political scientists notwithstanding, though, too few research studies of planning and policy analysis have explored these challenges as they arise in the practical conversations, the public-private negotiations, and the myriad meetings taking place daily in public agencies, private firms, or community neighborhoods. Although recent work on planning has begun to address the moral and political issues arising in ordinary practice (Hoch 1994, Healey 1997, Innes 1996, Marris 1990), the neglect of these issues in the broader policy studies community is all the more surprising because planners and policy analysts must often *do* this work on a daily basis, for better or worse: they must try to learn about appropriate technique, about what really matters here, and about other parties all at the same time (Fischer and Forester 1993, Storper and Sayer 1997).

Too often scholarly work on public learning stops where it should really begin: by framing the problems that hamper public deliberations. The professional culture of the social sciences unfortunately encourages this shortsightedness. Researchers shun the pragmatic, prescriptive work of recommending *what ought to be done,* in which sorts of situations. But planning and policy practitioners have to go where their academic colleagues have been trained not to go: into the prescriptive world of making good judgments about what to do. So while some social scientists take comfortable refuge behind the shamefully thin facade of a "value-free" social science, planning practitioners have to place their bets, as we have seen, staking their jobs and the welfare of affected citizens on those judgments at the same time (Forester 1999a,b).

What can be practical, though, about these stories that have not told us simply and directly "how to do it"? Precisely this: these stories provide a complexity and specificity that teach judgment, enrich perception, and heighten sensitivity. Most of all, these practice stories call our attention through the details to the kinds of considerations we need to take into account in our own novel circumstances. Practitioners and readers should rarely apply recipe-like instructions mechanically in changing circumstances. Instead, they can learn how to probe and improvise, to fit and adapt practical strategies to the complex situations they face.

Insightful practitioners' stories teach us not "for once and for all," despite contexts, what to do, but instead, in a range of complex circumstances, how to inquire, how to learn and go on, how to go about the tasks at hand. So, in chapters 1 and 2, we came to see how planners who do not take others' stories seriously are in danger of making themselves needlessly stupid, to say nothing of making themselves resented and feared. In chapter 3, as we considered accounts of practice from Israel and Norway, Arie Rahamimoff taught us lessons about consensus-based urban design, and we learned too how "the sketch" can serve as a ritual object to prompt inquiry.

In chapter 4, Rolf Jensen taught us about finding collaborative opportunities within otherwise adversarial negotiations, in his case with no less than the skeptical leadership of other city agencies. Baruch Hirschberg taught us about the prerequisites of the "diplomatic recognition" needed when environmentalists and transportation planners, among many others with deeply differing points of view, confront each other. So, too, then, in chapter 5 Ken Reardon, Silverio Gonsalez, Morten Leven, and Michelle Fine taught us about the challenges of community planning built from the bottom up, community planning based on participatory action research in neighborhoods of extreme poverty and weak public governance—in rural Venezuela and Norway, urban East St. Louis and a large East Coast city.

Chapters 6 and 7 explored several common challenges as they arise in deliberative practice: for example, the importance of "safe space," allowing parties not only to raise their serious concerns but also to work together seriously and creatively, collaboratively and adversarially, to resolve the issues before them as best they can; the role of planners acting as third parties, as de facto, not formal, mediators or convenors or facilitators or even catalysts working with diverse citizens to articulate and pursue their interests. If chapter 6 dealt relatively more with interests that stakeholders might bring to negotiations together, chapter 7 examined deliberative encounters taking place in the context of brutal histories of sexual domination and racism.

Chapter 8 made the underlying ethical issues more explicit than earlier chapters had, perhaps, and here we learned how many value issues come into play in even quite ordinary planning situations. We learned, too, the importance of moral improvisation—not the work of "anything goes," but the morally and practically demanding work that Martha Nussbaum has characterized as being richly responsible to general principles and obligations, while being keenly and practically aware of the particulars that matter in the unique situations at hand. We saw once more in this discussion that planners and policy analysts who fail to learn about value in real cases are likely to fail more generally. Missing what matters in the case, they may be sadly irrelevant at best, blindly destructive at worst.

Throughout these chapters we have seen that while reflective practitioners learn as they *act on practical situations,* deliberative practitioners learn as they *act with others* in the practical situations at hand. If planning is not to be the heavy-handed imposition of a small group's benign (or malign) decisions, it must foster not only voice and participation, but real learning, if not also practical deliberative decision making or the crafting of mediated multiparty agreements.

By exploring cases of deliberative practice through the accounts of practitioners themselves, we have indulged a bias for pragmatic hope, not wishful thinking, a bias for considering seriously the possibilities that lie before planners and public policy analysts, a bias for listening closely to practitioners while considering relevant theoretical literatures too. This book presents no last word on deliberation, democratic planning, or participation. Its purpose instead has been to integrate practical reflection with relevant commentary, so that those who wish to develop deliberative and participatory approaches to planning and policy analysis can move more easily ahead, building on the strengths of the practitioners in these chapters and dealing still more successfully with their weaknesses.

Notes

Introduction

1. These chapters complement and build on but do not repeat related studies of what planners do. See, e.g., the important work of Baum (1997a), Moore (1995), Hoch (1994), Innes (1995b, 1996), Healey (1997), Throgmorton (1996), Hendler (1995), Mandelbaum (1996), Howe (1994), Krumholz and Clavel (1994), and Riccucci (1995). Cf. Krumholz and Forester (1990). See note 7 as well.

2. Over the years, I have found that professional students treat discussions of potential benefits as interesting but hypothetical ideals, but those same students treat discussions of threats of failure more practically, less skeptically, and far more dramatically (Kahneman and Tversky 1979; cf. Cobb and Kuklinski 1997).

3. Such learning about ends arises inevitably in professional school curricula, even if social scientists consider the notion of "prescriptive knowledge" to be something of a contradiction in terms. Cf. Dyckman (1960) on the recognition of "ends" as an object of rationality in planning: "In an important political sense, planning is concerned with the rationality of ends" (p. 29). Nevertheless, Dyckman notes, "For the most part, planning has striven to test out this rationality-oriented drive on widely accepted ends. The selection of the physical environment as the major focus for this kind of control is more easily legitimatized than other kinds of planning. This tactic has permitted planning to evade open conflict of a sort that would inevitably raise the philosophical issues which we are investigating here. By restricting itself to a kind of therapy for the physical ills of the city, American city planning has been able to achieve widespread support from groups whose views on the general appropriateness of planning in society are widely divergent. But this has also prevented planning from facing the issue of its own mandate as squarely as it might" (pp. 30–31).

4. For exceptions, see the work of Hoch (1994), Healey (1997), and Innes (1996). Although recent work in political science and policy studies refers to such learning as "public deliberation," this label covers a variety of practices (and sins). Public deliberation can easily refer to discussions in Congress about legislative proposals (Bessette 1994, Mansbridge 1992), to state-level policymaking (Moore

1995, Innes 1994), to urban policy proposals and city governance (Hoch, Healey), as well as to more diffuse discussions in the public sphere such as PTAs, workplaces, and school boards (Benhabib 1996, Dryzek 1990). We focus here on the face-to-face work of public learning that constitutes the shared core of deliberation in these settings.

5. When planners simply try to preempt and control discussion, public deliberation is a sham; we should not confuse exercises in public manipulation and deception with serious efforts to promote public participation, learning, and deliberation. Such preemptive work of planners and policy advisers is only one among many political strategies of bureaucratic politics, itself a long-studied problem (e.g., Arnstein 1969, Lukes 1974, O'Connor 1973). If my *Planning in the Face of Power* paid more attention to the anticipation of, and response to, conditions fostering manipulation and misinformation (by others and planners too), this book is more concerned with clarifying the nature and possibilities (if not the probabilities) of effective democratic deliberative practices.

6. Cf. here Henri Frankfurt's definition of "bullshit" as "a lack of concern with the truth" (in Margalit 1993).

7. The scientific quality or objectivity of these chapters grows from their corroboration of the work of researchers such as Howell Baum (1990, 1997a), Martha Feldman (1989), Patsy Healey (1992, 1993a, 1996, 1997), Charles Hoch (1988, 1994), Judith Innes (1995a, 1995b, 1996), James March (1988), Peter Marris (1975, 1982, 1996), Mark Moore (1995), Lisa Peattie (1987), Robert Reich (1988), Tore Sager (1994a), Lynda Schneekloth and Robert Shibley (1995), and James Throgmorton (1996).

8. Even as some academics indulge in intellectual samurai encounters—with a slice of the verbose sword denouncing liberal failures, with another slice of correct politics taking "sides" with the invisible, the sub-altern, the excluded, the voiceless—planners have to make choices: how much risk to take, whose interests to defend and keep in the public eye? Worried about job security and persuaded by trickle-down promises, too many planners worry more about monied downtown interests than they do the urban poor, as Scott Campbell and Susan Fainstein (1996), Oren Yiftachel (1995), and Bent Flyvbjerg (1998) in their sanguine ways have taught us. Planners are no more noble human beings than lawyers, kindergarten teachers, or accountants. They embody no less ideology than any other professional or semiprofessional group of workers. They make mistakes. They succumb to temptations. They protect their own jobs. But many are nevertheless attracted to the vocation of planning because they wish to serve public need—and that desire deserves respect, recognition, and as importantly, political protection and encouragement.

9. Once we agree that we need to study political-economic structures and "forces," we may be more rather than less able to understand that in real spaces in real time, citizens have to figure out how to talk to (and write for) one another, and they have to do so in a world of racial suspicion, political uncertainty, fluid community ties and an ever shifting economy. This is a matter not simply of literal

communication, but of real performance, not of "mere words" but of deeds, and thus not of disembodied texts but of enacted political drama and tragedy.

Chapter 1

1. This chapter introduces the challenges of public deliberation (Dryzek 1990, Fischer 1995, Mathews 1994, Hoch 1994, Yankelovich 1991) by exploring the practice and politics of listening (Forester 1989) and the work that practice stories do. For related studies of participation in planning processes, see Healey (1997), Throgmorton (1996), and, on participatory budgeting in Brazil, Abers (1998).

2. Szanton (1981:159–160) summarized his results this way: "Third-party funders of [policy and program] advice (and especially federal agencies) tend to seek not merely useful truths, but useful truths of general applicability. They expect in this way to maximize return on their investment. Consultants suffer from the same temptation. . . . The intention is reasonable, but the results are poor. All communities believe themselves special, indeed unique. They want their advisors to address their *particular* concerns, not the problems of some category of communities to which a federal agency assigns them. The result is that where third-party funders insist on work whose results will be 'generalizable', city agencies lose interest, fail to cooperate, or flatly resist. . . . And Fitzgerald's irony holds: solutions to the problem of a particular city do prove useful elsewhere. Many urban problems are widely shared. Good solutions, therefore, do have wide potential. And urban officials across the country are linked by a profusion of professional associations . . . most of which meet regularly on national, regional, and statewide bases, and which also publish journals. News of useful innovations is thus conveyed in the least threatening and most convincing way—by the reports of fellow professionals. . . . 'Generalizability' will come, don't strain for it."

3. For an instructive account of objectivity as "critical intersubjectivity," see Fay (1996).

4. Donald Schön has shown us how practitioners "reflect in action" as they make moves, evaluate the results of those moves, and reconsider the working theories that guided those moves—as they consider what to do next. But the learning processes Schön focused on presume a good deal of practical knowledge on the practitioner's part. Before "moves" can be refined—lay the school out this way or that way; expand the program this way or that way—the practitioner needs to have taken a role in an institutional and political world. Not only are the roles typically ambiguous, but the political world is fluid as well. We must explore how planners learn as they listen, and significantly too, learn far more than "the facts" (cf. Schön 1983, 1990; chapter 8 below).

5. Just how much is to be attended to, probed, not missed, responded to sensitively, appreciated as significant in these stories is not typically obvious at all. Because these profiles and practice stories show us the vulnerability of planners' best-laid plans to much larger forces beyond their control, they echo the themes

of the classical tragedies. As political theorists and ethicists explore literature and theater as rich and vital sources of ethical and political teaching, we turn here to rich and vivid practice stories to explore the possibilities of ethically sensitive and politically astute planning practice. See especially here the work of Nussbaum and Euben.

6. Cf. Hannah Arendt (1971:445–446): "[Thinking] does not create values, it will not find out, once and for all, what 'the' good is, and it does not confirm but rather dissolves accepted rules of conduct. . . . The purging element in thinking, Socrates' midwifery, that brings out the implications of unexamined opinions and thereby destroys them—values, doctrines, theories, and even convictions—is political by implication. For this destruction has a liberating effect on another human faculty, the faculty of judgment, which one may call, with some justification, the most political of man's mental abilities. It is the faculty to judge *particulars* without subsuming them under those general rules which can be taught and learned until they grow into habits that can be replaced by other habits and rules."

7. Sandel (1982:181) writes, "Where seeking my good is bound up with exploring my identity and interpreting my life history, the knowledge I seek is less transparent to me and less opaque to others. Friendship becomes a way of knowing as well as liking. Uncertain which path to take, I consult a friend who knows me well, and together we deliberate, offering and assessing by turns competing descriptions of the person I am, and of the alternatives I face as they bear on my identity. To take seriously such deliberation is to allow that my friend may grasp something I have missed, may offer a more adequate account of the way my identity is engaged in the alternatives before me." We return to consider Sandel's insight again in chapter 6.

8. Cf. the ambiguities in the notions of *befindlichkeit,* care, and situatedness in Heidegger (1962) and more recent notions of care and relationship (Belenchy et al., 1986, Gilligan 1982, Tronto 1987).

9. It would be productive to generate the further synonyms here: mapping the ways we know we can make judgments that we would want friends to help us to reconsider (Pitkin 1972). Cf. discussions of self-command (Thomas Schelling 1984 in decision theory) and weakness of will (Rorty 1988, Nussbaum 1990). Can we read or listen to practice stories without sliding from "understanding" them to "accepting" them at face value, thus losing any ability to be critical of the practice they (re)present? Again, thinking about how we listen to friends may be helpful, for we seem no more to agree with or accept blindly whatever a friend says than we expect a friend blindly to agree with or accept whatever we (again, rashly, blindly, mistakenly) say. Luigi Mazza (personal correspondence) has raised a closely related question about the focus on the "micropolitics" of planning in Forester (1989). Cf. O'Neill (1972).

10. Chapter 2 continues to explore the subtle work of problem identification.

11. Commentators are careful to show that Aristotle's conception of friendship as an ethical ideal does not compromise broader claims of justice (Irwin 1989).

And just in the same way, I want to claim that we can learn from the stories of planners in ways that speak to the possibilities of justice, that do not compromise justice for the interests of a particular relationship we have.

12. For the most lucid discussions of the relationships of story and literature to deliberation, ethics and practice, see Martha Nussbaum's *Love's Knowledge* (1990). In a passage that could serve as the epigraph for this book, Nussbaum (1986:46–47) writes, "[The Greek tragedies] show us . . . the men and women of [the] Choruses making themselves look, notice, respond and remember, cultivating responsiveness by working through the memory of these events . . . and their patient work, even years later, on the story of that action reminds us that responsive attention to these complexities is a job that practical rationality can, and should, undertake to perform; and that this job of rationality claims more from the agent than the exercise of reason or intellect, narrowly conceived. We see thought and feeling working together, . . . a two-way interchange of illumination and cultivation working between emotions and thoughts; we see feelings prepared by memory and deliberation, learning brought about through *pathos*. (*At the same time we ourselves, if we are good spectators, will find this complex interaction in our own responses.*)" (emphasis added).

13. How can the messiness of case histories and profiles be an important part of their message? After all, planners face enormous social problems. How can they—and we—learn about good practice through messiness, complexity, and particular detail rather than general rules, universal maxims, and all-purpose techniques? Nussbaum (1986:186) recognizes clearly the suspicion that meets the suggestion, and even the tradition, "that defends the role of poetic or 'literary' texts in moral learning." Taking this suspicion head on, she writes, "Certain truths about human experience can best be learned by living them in their particularity. Nor can this particularity be grasped solely by thought 'itself by itself.' " . . . It frequently needs to be apprehended through the cognitive activity of imagination, emotions, even appetitive feelings: through putting oneself inside a problem and feeling it. But we cannot all live, in our own overt activities, through all that we ought to know in order to live well. Here literature, with its stories and images, enters in as an extension of our experience, encouraging us to develop and understand our cognitive/emotional responses."

14. C. W. Churchman once described in class the pragmatist's theory of truth as follows: A "is" B (the car is red, the housing is substandard, it is raining) is to be read, "A 'ought to be taken as' B"—thus taking descriptions to be pragmatic and selective actions, not correspondence-like statements picturing a brute reality. (Churchman 1971; Murdoch 1970; Pitkin 1972.)

15. LM to PH, personal correspondence, June 22, 1990.

Chapter 2

1. This is not to argue that stories are always better than studies. Stories can be more or less relevant, as studies can be, in any given case. Stories may be better or worse (along many criteria), as studies can be; both may be shown to be riddled

with falsehood or not, to attribute causality or responsibility faultily, to be selective in justifiable or unjustifiable ways. To blur boundaries, stories may make reference to studies, and studies may refer to or even share characteristics of stories in their own right. Cf. Gusfield (1981), Van Maanen (1988), Abrams (1991), Hummel (1991). In addition, stories partially constitute their narrators, as well as being told and improvised by them (Forester 1993b).

2. Practice stories can teach us about the priority of practical rationality: before problem solution proceeds, problem construction must be well underway. For the argument applied to medical practice and medical ethics, see Jennings (1990). Cf. George (1994), Forester (1993a), and Nussbaum (1990:chap. 2). As we shall see, these practice stories can teach listeners what it may be like to face—in Nussbaum's terms, to "be finely aware and richly responsible to"—the issues they portray (Nussbaum 1990, chap. 5). Such stories provide moral phenomenologies that more abstract philosophical or structuralist accounts do not.

3. Of course, emotions can mislead us, just as some selections of "the facts" can. Nevertheless, failing to respond with emotional sensitivity can lead to as much trouble as failing to provide "the facts." Cf. Nussbaum (1990a:75–82). On the place of such emotional sensitivity in acting well, in rational action, see Nussbaum (1986, p. 364): "Aristotelian ethical knowledge . . . consists, above all, in the intuitive perception of complex particulars. Universals are never more than guides to and summaries of these concrete perceptions; and 'the decision rests with perception'. *Perception, furthermore, is both cognitive and affective at the same time: it consists in the ability to single out the ethically salient features of the particular matter at hand; and frequently this recognition is accomplished by and in appropriate emotional response as much as through intellectual judgment*" (emphasis added). We must try to recognize difference and listen carefully, presuming neither that differences of experience, class, gender, or race, for example, must be unbridgeable and mutually incomprehensible nor that some perfect intersubjectivity will ensure equally perfect understanding.

4. Holding aside disputes about true and false stories, any one person might tell an infinite number of different (!) true stories about the same event—from different angles, with different emphases, with different amounts of detail, for example. We should not confuse the plurality of possible stories with simplistic notions of truth or falsity. We should assess the ways practice stories might be relevant or irrelevant, sensitive or callous, simplistic or nuanced, timely or not, confusing or edifying, and so on. Cf. John Austin's (1961:131) remark: "If only we could forget for a while about the beautiful and get down instead to the dainty and the dumpy."

5. To say that institutional pressures shape the stories that are told is true but hardly all determining. Cf. the warning of Bruce Jennings (1990:269, emphasis added): "What ethicists do not do is see moral concepts and categories as *embedded* in ongoing forms of social practice and experience that are *structured* via particular institutional patterns or the encounter with certain technological constraints. *And ethicists do not pay much attention to the ways in which struggling*

with a problem or acting within a certain pattern of constraints or power relationships can actually transform the moral perception and understanding of agents" (cf. Forester 1989, 1993b). Planners' stories too may be hegemonic or counterhegemonic, and we need better to assess which they are when.

Chapter 3

1. I am indebted to the generosity of Rolf Jensen and Arie Rahamimoff for the accounts of their practice excerpted here from fuller profiles (Forester 1994a; Forester, Fischler, and Shmueli 1997).

Chapter 4

1. Practice can lead theory, and in planning, the practice of astute, sensitive, and skillful planners can sometimes lead the more abstracted theories of planning academics—or that, at least, is the suggestion that underlies the argument that follows. I am indebted to the generosity of David Best, Rolf Jensen, and Baruch Hirschberg for the accounts of their practice excerpted here from taped interviews and fuller profiles (Forester 1994a; Forester, Fischler, and Shmueli, 1997).

2. Miller (1992). See also Nussbaum (1990a: especially chap. 2). See chapter 7 in this volume.

3. Because all three of these professionals happen to be men, this chapter will use the pronoun *he* throughout. The argument, however, is not limited to men; it echoes recent feminist literature on the importance of relationships, "difference," and listening (e.g., Young (1990), Benhabib (1992), and earlier, Gilligan (1982)).

4. We seem often to mistake another's adversarial intentions for the actual constraints of the situation. This is a bit like assuming that a used car dealer's first price is her last and best. For an account of deliberation within adversarial contexts, see Bessette (1994), Mathews (1994), and Mansbridge (1992). Cf. Esquith (1994:91–92) on the political judgment required within democratic deliberation: "Political judgment—that combination of character and practical wisdom that prepares citizens to perceive power relations and respond to them justly together—is not the judgment of rulers, it is the judgment of participants."

5. In any negotiation, "creating value" and "claiming value" exist in an uneasy political tension. "Creating" value may expand the pie to be "claimed," but "claiming" prematurely—getting yours, out of fear, defensiveness, or distrust—can prevent any "creating" at all and produce mediocre outcomes for everyone (Lax and Sebenius 1990).

6. David Mathews (1994:112) makes the more general, less institutionally specific point: "Whether representative or direct, local or national, democracies require public deliberation. Deliberative politics is not the ideal politics; it is the

necessary politics of a democracy." One can argue this much whether one imagines communitarian or more argumentative, discourse-theoretic, forms of deliberation (cf. Barber 1984, Benhabib 1992, chapter 6 below; cf. Habermas 1996: 284.

7. Here Hoch shows us that deliberation is not something we can "forget" as an overly idealistic, and potentially conservative, idealization of political discourse, as some political scientists might suggest (Sanders 1997). In the abstract, calls for generalized deliberations can too easily be blind to the historical and institutional impediments to such political conversations; racism, sexism, class exclusion, problems of scale and access, and others (Sandercock and Forsyth 1992). But Hoch makes no abstract and general case for deliberation. He shows us that planners and public managers have no choice but to foster—or squander!—opportunities for deliberation in their everyday working meetings. The situated and entangled character of planning and administrative practice forces attention to deliberation, and only those free of political interdependence can pretend to forget it.

8. At a recent planning staff meeting I attended in a small municipality, the question of structuring community participation to allocate funds designated for street improvements arose in the context of a poor neighborhood in which the staff could already hear, "We don't need street improvements; we need better housing." Cf. Elkin (1987).

9. These accounts should be read in the light of recent literature on deliberation in law (Michaelman 1988, Sunstein 1988), political theory (Barber 1984, Dryzek 1990, 1995, Fischer and Forester 1993, Fischer 1995, Yankelovich 1991), philosophy (Cohen 1993, Gutmann 1993b, Habermas 1996, Nussbaum 1990a, Putnam 1990), and planning and design (Forester 1995, Schneekloth and Shibley 1995). The practitioners' accounts that follow tell us a good deal more about the conduct of public deliberations than their design. Further work must explore the questions of institutional design more systematically (cf. Dryzek 1995).

10. Lawrence Susskind gives a fascinating and similar account of joint fact finding involving engineers' and citizens' projections of physical scale in a transportation-environment land use controversy (in Forester 1994b).

11. For an important discussion of the ways we learn about value as well as fact, see Nussbaum (1990a) and cf. chapter 8, this volume.

12. For a striking argument that suggests in skeletal form how deliberation shapes parties' preferences, recognition of one another, and subsequent arguments, see David Miller (1992:55). Cf. Taylor (1998).

13. Cf. Nussbaum (1990a:74): "The content of rational choice must be supplied by nothing less messy than experience and stories of experience. Among stories of conduct, the most true and informative will be works of literature, biography, and history; the more abstract the story gets, the less rational it is to use it as one's only guide. Good deliberation is like theatrical or musical improvisation, where what counts is flexibility, responsiveness, and openness to the external; to

rely on an algorithm here is not only insufficient, it is a sign of immaturity and weakness."

14. Cf. Fishkin (1991:29) on the significance of democratic deliberation: "Without deliberation, democratic choices are not exercised in a meaningful way. If the preferences that determine the results of democratic procedures are unreflective or ignorant, then they lose their claim to political authority over us."

15. I tape-recorded my interview with Baruch Hirschberg in Haifa, Israel, on October 25, 1993.

16. Notice that this is what Jensen did by going "one level down" to the harbor engineer, shifting the discussion from the deadlocked agency directors to a "parallel negotiating track" between staff. The term is Larry Susskind's (personal conversation, October 1995).

17. Compare Hoch's conclusion (1994:337): "Professional planners work in an institutional order of competitive and hierarchical relationships, which, despite their adversarial and instrumental qualities, require some cooperation. Many professional planners regularly try, in imaginative, incremental, and occasionally grand ways, *to shift attention from the adversarial to the deliberative*. Their persistence testifies to the effort they are making in our competitive liberal society to keep alive the practical possibilities and the hope of responsible, free, and informed deliberation."

Chapter 5

1. I wrote this chapter to sort through the perplexity I felt after listening to years of fascinating discussions of participatory action research (PAR) at Cornell University. Both practitioners' and academics' analyses of PAR seemed far narrower than their own stories of deliberative learning and political rationality seemed to deserve.

2. The research was reported internationally in a paper to the XII World Congress of Sociology in 1990, and in an article in the *Journal of Social Issues,* V46, 1990, authored by Briceno, Gonzalez, Phelan, and Ruiz, Laboratorio de Investigaciones Sociales, Universidad Central de Venezuela, 1987.

3. Levin (1993) characterizes action research in the following terms: "According to [the cogenerative model of AR described by Elden and Levin (1990)], AR involves a mutual learning process that generates new local theory. Local theory is a shared social construction between insiders and outsiders relevant to the problem in focus that emerges from interaction in common social settings. The construction of this new local theory is molded through interaction in many fora created by the action research process. . . . The process begins with problem definition. Real life challenges are addressed. . . . Insiders, the problem owners, 'cogenerate' with the outside researcher an identification of the problem. This is accomplished through dialogues. . . .

"According to this model [of cogeneration], the essence of action research is the creation of a new social reality through discourse that encourages and supports learning. The co-generative learning process is based on 'democratic dialogues' (Gustavsen 1992; Habermas 1984). A true dialogical situation must be free of dominance by outsiders. . . . The co-generative process implies equality in the sense that every argument is valid and must be dealt with to the degree that it is fruitful in creating new relevant knowledge." (Cf. Elden and Levin, 1990.)

4. Commenting on the AR process as he saw it, Levin writes in an earlier passage: "Learning from the AR process occurs in at least three ways. First, learning may occur directly during the democratic dialogues. Learning also emerges from reflection on the outcomes of different actions. Third, the experience of writing about outcomes creates new understanding in a different and more structured system. . . . In the research process, both accomplishment of goals and the reflection process involves intensive group interaction. The collective action itself becomes a learning experience. Therefore, the outsider's action competency must contain at least two elements. First, the researcher who uses the cogenerative AR approach must design and manage fora for effective social exchange. The researcher must also be able to help develop knowledge based on social science criteria for knowledge production" (1993:198).

5. On the consideration of fundamental disagreements in deliberative processes, see Amy Gutmann (1993a:199) who writes, "Deliberation may sometimes increase moral conflict in politics by opening up forums for argument that were previously closed. . . . Deliberation encourages people with conflicting perspectives to understand each other's point of view, to minimize their moral disagreements, and to search for common ground, but it begins by opening politics up to a range of reasonable disagreement that is restricted by less deliberative politics."

6. For recent theoretical work on dialogue still concerned with epistemology more than with ethics, with meaning and assumptions more than with speech and performance, with the nature of underlying paradigms more than with rituals of contingent promising, see Isaacs (1993). On laboratories for judgment, Peter Koschitz writes aptly that "by assessing the possible outcomes of a measure meant to solve a certain problem we can learn much more about what we value than by theoretically discussing value priorities" (personal correspondence, September 16, 1993, Swiss Federal Institute of Technology, Zurich).

7. Personal conversation, Mimrebu seminar, Norway, October 1993.

8. Once we recognize how our own limited attention constrains us, we can then begin to realize that our decision-making processes can aggravate or correct, compound or remedy our limited abilities to imagine the future, to know, when we need to, what is relevant and what is not, what is significant to this case and what is not.

9. Myerhoff (1982:111) refers to special moments of recollection in sharing stories: "To signify this special type of recollection, the term, 'Re-membering' may be used, calling attention to the reaggregation of members, the figures who belong

to one's life story, one's own prior selves, as well as significant others who are part of the story. Re-membering, then, is a purposive, significant unification, quite different from the passive, continuous fragmentary flickering of images and feelings that accompany other activities in the normal flow of consciousness. . . . A life is given a shape that extends back in the past and forward into the future. It becomes a tidy edited tale. Completeness is sacrificed for moral and aesthetic purposes. Here *history may approach art and ritual.* The same impulse for order informs them all" (emphasis added).

10. Attending to objects, we can reconsider roles and our own possibilities of action in the world of those objects: the soccer ball (I can play just a bit), the car (I have little time or money to travel), the radio (I enjoy, and look forward to, listening to the blues) and so on. Rituals give us stories; stories give us significant objects; significant objects give us regulative and constitutive norms and roles; norms and roles give us ways of going on separately or together, that is, fluid social identities; identities give us a sense of possibility (and limits) and relevant questions to explore regarding related concerns (Could I get time off to travel with the car?) . . .

11. There is much more to be said here. Storytelling is performance, and the style of the performance can be as significant as any content of the story. Most important, value is articulated and allocated not so much in the story but in the action of its telling: the warning or encouraging or challenging or reminding or reassuring that the story*telling* achieves as practical-allocative communicative action (Forester 1991b).

12. This helps us to see why people often say they "love" or "hate" ritual. In both cases, ritual participation is understood to shape identity. Those who "hate ritual" understand that their own ritual participation presumes to make them something, someone, they do not want to be. Ritual haters often express their dislike of hypocrisy: that the identity affirmed and presented in the ritual performance is not genuine. Those who "love ritual" often find ritual participation a means of honoring what they cannot express, a means of being part of a larger community of practitioners with a history of beliefs, aspirations, values, and commitments that again finds poorer expression in ordinary language than in the pomp and circumstance of highly sedimented, historically resonant ritual practices. All of which is not to make a case for or against any particular ritual but to appreciate that ritual performance shapes social identity, which both lovers and haters of ritual understand. In general, if we value something, we want to participate in the rituals honoring and attending to it. Conversely, we resist participating in rituals associated with what we do not honor because the ritual performance makes us someone we are not, makes us duplicitous, and often publicly so.

13. In public policy settings we often find the rituals called "formal public hearings." These are clearly structured, even deliberately stilted forms of public presentation that channel the stories of participants to a formal hearing board and

prevent participants from conversing and forming any joint agreements among themselves. Such public hearing rituals often seem to exacerbate conflict because they encourage exaggeration and posturing while they discourage clarification of issues, testing of claims, and exploration of options or potential agreements. These ritualized hearings constitute their participants not as citizens who must speak and come to terms together but as poker-playing adversaries, as masters of the bluff rather than as good listeners able to respond to one another.

14. The emotional texture of PAR and deliberative processes appears to be a seriously neglected area of inquiry. Cf. Nussbaum (1990a:81): "It frequently happens that theoretical people, proud of their intellectual abilities and confident in their possession of techniques for the solution of practical problems, are led by their theoretical commitments to become inattentive to the concrete responses of emotion and imagination that would be essential constituents of correct perception." On the importance of emotion in deliberation, cf. Dewey (1930).

15. Cf. Fishkin (1991:29): "Deliberation is necessary if the claims to democracy are not to be de-legitimated."

16. Again we learn about such value allocations by evaluating others' ritual performances: so someone means to help but does so ineptly, or callously (bedside manner), and we see that in addition to intention, value is allocated. By attending to ritual performance, we can not only respect (more or less) and recognize the person, but we can assess the way he or she shapes or obscures or responds to or illuminates or threatens parts of the social world we share (particular relationships, or next week's meeting).

17. We shape the allocation of value in the world in ways we do not intend, and we can view others' actions as similarly shaping value in nonintentional ways. So we say, for example: (1) "I know you didn't mean to hurt her feelings, but you did" (regarding the value of another's integrity), or (2) "I know you didn't mean to disclose the secret" (regarding the value of particular information in a relationship), or (3) "I know you didn't mean to get us into this trouble" (regarding the value of a violated norm), or (4) "I know you didn't mean to break the computer" (regarding the value of a valued object), and so on. This suggests that we commonly evaluate our own and others' actions as they affect value in the world beyond whatever we or others intend.

18. There are issues here of tragic limits of attention in complex decision situations. We may care about far more than we attend to at the beginning of a negotiation, for example, and the complexity of decision making and participatory rituals may enable us to acknowledge more (or less, depending on their form and content) of those concerns. Cf. Nussbaum (1986:chap. 2) and its related arguments against the neglect of loss encouraged by facile beliefs in the commensurability of value concerns. The leader who ignores the losses caused by his or her "correct" decisions may be not only blind but blinding, neglecting important values and leading others to fail to appreciate them as well (cf. Davy 1997).

Chapter 6

1. This chapter complements chapter 5's exploration of participatory action research. As we shall see, mediated negotiation strategies present citizens with several of the promised benefits of PAR: greater participation in affairs affecting their lives, greater ownership of results, more effective crafting of solutions as a result of using local knowledge, "empowerment," and so on. Substantial as they are, promises are one thing, practice another. To deliver the promises, citizens in a wide variety of settings will need to understand better the complex character of public dispute-resolution processes and mediated negotiations, our focus here, in particular.

2. My research assistant obtained these (among others) in the mail after telephoning roughly a dozen suppliers of public dispute-resolution services and asking for their materials. We identified these suppliers by consulting the "Resources" section of *Consensus*, a recent glossy bringing word of dispute-resolution practice and issues to roughly 25,000 public officials.

3. Material quoted was obtained by writing: New England Environmental Mediation Center, 30 Sumner Street West, Gloucester, Massachusetts 01930-1548.

4. Material quoted was obtained by writing: ICF, 9300 Lee Highway, Fairfax, Virginia 22031.

5. Material quoted was obtained by writing: CDR Associates, 100 Arapahoe, Boulder, Colorado 80302.

6. Cf. Forester (1998a). Bill Potapchuk suggests that mediators' interpretations of their obligations to honor confidentiality significantly constrain their accounts of their own practice (letter to author, February 1990).

Chapter 7

1. Elster (1983) writes, "Endogenous preference change by learning not only creates no problems for ethics, but is positively required by it. If trying out something you believed you would not like makes you decide that you like it after all, then the latter preferences should be made the basis for the social choice, and social choice would not be adequate without this basis" (pp. 138–139). Regarding deliberate character planning, Elster suggests, "Whereas adaptive preferences typically take the form of downgrading the inaccessible options, deliberate character planning would tend to upgrade the accessible ones. . . . In a less than perfect marriage I may adapt either by stressing the defects of the wise and beautiful women who refused me, or by cultivating the good points of the one who finally accepted me. The latter adaptation, if I can bring it off, clearly is better in terms of cardinal want satisfaction" (p. 119). Further, "Character planning and adaptive preference formation have very different consequences for *freedom*. . . . It would hardly do to say that the free man is one whose wants have shrunk to the vanishing point out of sheer resignation, but there is a respectable and I believe valid

doctrine that explains freedom in terms of the ability to accept and embrace the inevitable" (p. 119).

2. Nussbaum (1990a): "Intellect will often want to consult these feelings to get information about the true nature of the situation. Without them, its approach to a new situation would be blind and obtuse. . . . Perception is not merely aided by emotion but is also in part constituted by appropriate response. Good perception is a full recognition or acknowledgement of the nature of the practical situation; the whole personality sees it for what it is. The agent who discerns intellectually that a friend is in need or that a loved one has died, but who fails to respond to these facts with appropriate sympathy or grief, clearly lacks a part of Aristotelian virtue. It seems right to say, in addition, that a part of discernment or perception is lacking. This person doesn't really, or doesn't fully, *see* what has happened, doesn't recognize it in a full-blooded way or take it in. We want to say that she is merely saying the words, "He needs my help," or "she is dead," but really doesn't yet fully *know* it, because the emotional part of cognition is lacking. . . . Neither is it just that the emotions supply extra praiseworthy elements external to cognition but without which virtue is incomplete. The emotions are themselves modes of vision, or recognition. Their responses are part of what knowing, that is truly recognizing or acknowledging, consists in" (p. 79).

Nussbaum weaves together here the cognitive and the performative work of recognition, but we might try to tease these epistemological and ethical moments apart—to ask how, for example, to encourage each capacity.

3. The preceding lines read: "If we define practical issues as issues of the good life, which invariably deal with the totality of a particular form of life or the totality of an individual life history, then ethical formalism is incisive in the literal sense: the universalization principle acts like a knife that makes razor-sharp cuts between evaluative statements and strictly normative ones, between the good and the just" (Habermas 1990:104). Cf. Seyla Benhabib, who distinguishes communicative ethics from "consent theories": "Consent alone can never be a criterion of anything, neither of truth nor of moral validity; rather, it is always the rationality of the procedure for attaining agreement which is of philosophical interest. We must interpret consent not as an end-goal but as a process for the cooperative generation of truth or validity. The core intuition behind modern universalizability procedures is not that everybody could or would agree to the same set of principles, but that these principles have been adopted as a result of a procedure, whether of moral reasoning or of public debate, which we are ready to deem "reasonable and fair." It is not the *result* of the process of moral judgment alone that counts but the process for the attainment of such judgment which plays a role in its validity and, I would say, moral worth" (Benhabib and Dallmayr, 1990, 345).

4. Habermas notes as much, but does not develop the point (Habermas 1990: 105–106). Cf. Young's (1990; chapter 2) distinguishing five faces of oppression: exploitation, marginalization, powerlessness, cultural imperialism, and violence.

5. Pursuing the issues of justifiability, Habermas continues, "Thus the development of the moral point of view goes hand in hand with a differentiation within

the practical into *moral questions* and *evaluative questions*. Moral questions can be decided rationally, i.e. in terms of *justice* or the generalizability of interests. Evaluative questions present themselves at the most general level as issues of the *good life* (or of self-realization); they are accessible to rational discussion only *within* the unproblematic horizon of a concrete historical form of life or the conduct of an individual life" (Habermas 1990:108).

6. Cf. Benhabib and Dallmayr (1990:349): "Since practical discourses do not theoretically predefine the domain of moral debate and since individuals do not have to abstract from their everyday attachments and beliefs when they begin argumentation, however, we can accept that not only matters of justice but those of the good life as well will become thematized in practical discourses. A model of communicative ethics, which views moral theory as a theory of argumentation, need not restrict itself to questions of justice. . . . It is crucial that we view our conceptions of the good life as matters about which intersubjective debate is possible, even if intersubjective consensus, let alone legislation, in these areas remains undesirable. However, *only through such argumentative processes can we draw the line between issues of justice and of the good life in an epistemically plausible manner, while rendering our conceptions of the good life accessible to moral reflection and moral transformation*" (emphasis added; note criticism of note 6).

7. Cf. Elster (1983) on adaptive preferences, and Minard, Jones, and Patterson (1993:4), evaluating the process of comparative risk assessment in six states: "Virtually everyone we interviewed said that the effort of ranking risks (or risk management priorities) forced the participants to deal with the data and their values in a powerful and productive way. Most projects have charged committees with the task of ranking problems, forcing people with diverse interests to come to terms with each other as well as with the data. The process demands critical thinking and forces decisions. It exposes weak arguments, poor data, and fuzzy thinking—all of which have been present in varying degrees in the state projects."

8. If all this were accepted for the sake of argument as true, Habermas's procedural testing in "discourse ethics" would still be untouched and still have its work to do. But before it can do that work, parties to the discourse must have done the work of recognition, learning about the other and themselves, and so perhaps have *stopped learning*. Analytically we might distinguish efforts of self- and other-recognition, with its associated learning about needs, interests, hopes, and aspirations, identity-constituting features and contingencies, from the testing of validity of norms of action, but thinking of them in sequence might be quite misleading if clarifications of a party's suffering or constitutive concerns recast our analysis of the generalizability of interests.

9. Benhabib writes, "Moral vision is a moral virtue, and moral blindness implies not necessarily an evil or unprincipled person, but one who cannot see the moral texture of the situation confronting him or her. Since the eighteenth century, ethical rationalism has promoted a form of moral blindness with respect to the moral experience and claims of women, children, and other "nonautonomous others," as well as rough handling the moral texture of the personal and familial"

(Benhabib and Dallmayr 1990:357). Cf. Iris Young: "One possible interpretation of communicative ethics is that normative claims are the outcome of the expression of needs, feelings, and desires which individuals claim to have met and recognized by others under conditions where all have an equal voice in the expression of their needs and desires. . . . A strong strain of Kantian universalism remains in Habermas, however, which undermines this move to a radically pluralist participatory politics of need interpretation" (Young 1990:118). Cf. Honneth and Wright (1995).

10. For the argument that "viewing justice as the center of morality unnecessarily restricts the domain of moral theory," see Seyla Benhabib (Benhabib and Dallmayr (1990:348–349). Benhabib argues that we should understand communicative ethics as continuous with ordinary moral conversation rather than in an idealized world: "Each time we say to a child, 'But what if other kids pushed you into the sand, how would you feel then?', and each time we say to a mate, or to a relative, 'But let me see if I understand your point correctly,' we are engaging in moral conversations of justification. And if I am correct that it is the process of such dialogue, conversation, and mutual understanding, and not consensus which is our goal, discourse theory can represent the moral point of view without having to invoke the fiction of the *homo economicus or homo politicus.*" (Cf. Benhabib 1992:52.)

11. Both historians and psychiatrists have noted the importance of the language in which trauma is represented, as we shall see. Judith Herman closes her introduction by writing, "I have tried to communicate my ideas in a language that preserves connections, a language that is faithful both to the dispassionate, reasoned traditions of my profession and to the passionate claims of people who have been violated and outraged. I have tried to find a language that can withstand the imperatives of doublethink and allows all of us to come a little closer to facing the unspeakable" (Herman 1992:4). Finding such a language, Benhabib, Luban, and Arendt make clear, is no small matter (Benhabib 1990, Luban 1983).

12. Current notions of deliberation certainly raise issues of exclusion (e.g., Barber 1984, Young 1990), but the important point is that they hardly seem to explore their own implications: the real challenges of inclusion. Until our political theories do that, we will continue to hear the sound of one hand clapping: deconstructive critiques without reconstructive political analyses.

13. Cf. Seyla Benhabib: "If we view discourses as a procedural model of conversations in which we exercise reversibility of perspectives either by actually listening to all involved or by representing to ourselves imaginatively the many perspectives of those involved, then this procedure is also an aspect of the skills of moral imagination and moral narrative which good judgment involves, whatever else it might involve. I do not therefore see a gulf between moral intuition guided by an egalitarian and universalist model of moral conversation and the exercise of contextual judgment" (Benhabib and Dallmayr 1990:363).

14. Cf. Santner (1992:153): "To take seriously Nazism and the 'Final Solution' as massive trauma means to shift one's theoretical, ethical, and political attention

to the psychic and social sites where individual and group identities are constituted, destroyed, and reconstructed. This mode of attention is one which, to paraphrase Freud, though it may not always contradict the pleasure principle, is nevertheless independent of it and is addressed to issues that are more primitive than the purpose, narrative or otherwise, of gaining pleasure and avoiding unpleasure. It is furthermore a mode of attention that requires a capacity and willingness to work through anxiety."

15. To note a striking parallel, consider that Martha Nussbaum (1986) criticizes assumptions of commensurability in decision making for hiding the particularity of loss from our view. Benefit-cost analysis, for example, might then be understood as a highly revelatory or psychologically defensive, blinding mechanism of Santner's narrative fetishism. On the way that justification and explanation individuate actions, see Rorty (1988:288).

16. LaCapra (1994:214) writes, "Any attempt to facilitate processes of mourning and to further the emergence of viable public rituals would require an effective critique of anti-Semitism and related forms of scapegoating and victimization. It would be a mistake to concentrate all attention on postwar Germany, for anti-Semitism and scapegoating as well as the more particular problem of working-through denial and repression with regard to the Holocaust arise in other countries, including the United States."

17. Cf. Judith Herman (1992:228, 231) suggesting ritualized group processes of working through trauma (his)stories: "Group members can help one another to bear the pain of mourning. The presence of other group members as witnesses makes it possible for each member to express grief that would be too overwhelming for a lone individual. As the group shares mourning, it simultaneously fosters the hope for new relationships. Groups lend a kind of formality and ritual solemnity to individual grief; they help the survivor at once to pay homage to her losses in the past and to repopulate her life in the present. . . . The rituals of sharing offer tangible reminders of present connections even as each survivor remembers her moment of being most alone."

18. The same is true for public dispute-resolution processes involving facilitation and mediated negotiations (Kelman 1992). For a closely related analysis of participatory research processes, see chapter 5. The need to act together has led Yehezkel Dror (1992) to reflect politically on the sense in which we may see "trauma as an opportunity." Dror's argument raises important ethical issues that we must address if we are to respond to trauma politically as well as therapeutically.

19. Perhaps no other work signals the significance of working-through processes for public deliberations as does Dan Bar-On's on meetings of children of Holocaust survivors with children of Nazi perpetrators (cf. Bar-On 1989, 1999). This chapter developed in part from a previous year's conversations with Dan Bar-On.

20. Cf. Herman (1992:220): [In groups devoted to creating a safe space for trauma survivors], "strong cohesion among group members is not required to create an atmosphere of safety; rather, the safety inheres in the rules of anonymity

and confidentiality and in the educational approach of the group." But should anyone think that liberal legal and administrative protections already effectively ensure "safety" in public deliberations, the voices of women and people of color should lead to reconsideration.

21. What does this mean for the study of democratic deliberation? Recall that LaCapra understands the historian's work to be in part practical, critically rigorous but nevertheless a worldly intervention facilitating or retarding working-through processes, and thus subject to "ritual as well as aesthetic criteria." Cf. David Luban (1983:248) on Hannah Arendt's similar struggle against the dangers of reductive theorizing. Cf. Benhabib (1990:186): "The moral resonance of one's language does not primarily reside in the explicit value judgments which an author may pass on the subject matter; rather such resonance must be an aspect of the narrative itself. The language of narration must match the moral quality of the narrated object. Of course, such ability to narrate makes the theorist into a storyteller, [even if] it is not the gift of every theorist to find the language of the true storyteller."

Chapter 8

1. For a recent profession-centered view of ethics in planning, see Howe (1994).

2. Henry Richardson (1990) develops an account of practical judgment that avoids the extremes of "application" (the rules are clear and well ordered; implications, given the case, follow straightforwardly) and "balancing" (rules conflict, but we balance them intuitively to determine proper action) by centering on the work of "specification" (given general goals of minimizing traffic delays and improving physical infrastructure, we could add "when necessary" to the latter, specifying it a bit, and so consider van pooling as an acceptable option satisfying both "minimize traffic delays" *and* our newly specified "improve infrastructure when necessary").

3. On a calculating view, the joy of victory might cancel out the agony of defeat (as if, in sum, nothing has happened), and for the saying, then, "You win some, you lose some," one might substitute, summing up, "Life doesn't come to much." On the view finding judgment central, though, the joys of victory and the agonies of defeat reshape the lives of victors and losers alike, and the saying, "You win some, you lose some," sketches the demands likely to be put on character—bearing victory without arrogance and defeat without resentment, for example. For a far-reaching analysis, see Sager (1994a) and references to Nussbaum's work in chapter 7.

4. I am interested here in moral realism. Kate must be concerned not only with citizens' expressions of values, their espoused beliefs, but also with the value of the conditions to which such beliefs refer. Cf. Iris Murdoch (1970) referring to "the realism which must involve a clear-eyed contemplation of the misery and evil of the world" (p. 61); and "the realism (ability to perceive reality) required

for goodness is a kind of intellectual ability to perceive what is true" (p. 66); cf. Hilary Putnam (1987). Cf. references in chapter 6 to the shaping of value.

5. Murdoch (1970:67), stressing the ability to see clearly, the freedom from fantasy, which she takes to be "the realism of compassion," writes, "It is what lies behind and in between actions and prompts them that is important, and it is this area which should be purified. By the time the moment of choice has arrived the quality of attention has probably determined the nature of the act." Amelie Rorty (1988:286) writes, "Emphasis on judicialism unfortunately tends to focus on the moment of radical, selective choice, rather than on the delicate and difficult processes that precede choice, the processes of generating and articulating the range of alternatives."

6. Such general improvisation we might call bargaining or incrementalism or even boundedly rational action—but theories of bargaining, incrementalism, and bounded rationality all presume a process of search for acceptable agreements within a given space of interests. When planners deliberate well, they alter that space and may then transform themselves and enable participants to grow (to be "empowered") in the process (Forester 1993b, 1999b).

7. Murdoch (1970:66) writes, "The authority of the good seems to us something necessary because the realism (ability to perceive reality) required for goodness is a kind of intellectual ability to perceive what is true, which is automatically at the same time a suppression of self. *The necessity of the good is then an aspect of the kind of necessity involved in any technique for exhibiting fact.*"

8. This account helps us see why ideologues may not be particularly practical people; they are more attentive to their own "line" than to the needs and circumstances of their audiences. To the counterargument holding that demagogues are ideologues who are extraordinarily and dangerously practical, we might say that a demagogue is a moral improviser of a decadent sort: a moral improviser who acts unilaterally, reductively, and manipulatively.

9. As Seyla Benhabib (1988:43) writes, her "moral judgment . . . certainly must involve the ability for "enlarged thought," or the ability to make up [her] mind "in an anticipated communication with others with whom I know I must finally come to some agreement." Arendt and Benhabib certainly do not mean by "others with whom I know I must finally come to some agreement" simply one's supervisor at work, or moral judgment would be reducible to pleasing one's superiors, no matter what. Such capacity for judgment, Benhabib (1988:44) continues, "to 'think from the perspective of everyone else,' is to know 'how to listen' to what the other is saying, or when the voices of others are absent, to imagine oneself in a conversation with the other as my dialogue partner. 'Enlarged thought' is best realized through a dialogic or discursive ethic."

References

Abers, Rebecca. 1998. "Learning Democratic Practice: Distributing Government Resources Through Popular Participation in Porto Alegre, Brazil." In M. Douglass and J. Friedmann, eds., *Cities for Citizens*. New York: Wiley.

Abrams, Kathryn. 1991. "Hearing the Call of Stories." *California Law Review,* 79(4):971–1052.

Alexander, Ernest. 1996. "After Rationality: Towards a Contingency Theory for Planning." In Seymour Mandelbaum, Luigi Mazza, and Robert Burchell, eds., *Explorations in Planning Theory*. New Brunswick, N.J.: Center for Urban Policy Research.

Alexander, Gregory. 1989. "Dilemmas of Group Autonomy: Residential Associations and Community." *Cornell Law Review* 75:1–61.

Amy, Doug. 1987. *The Politics of Environmental Mediation*. New York: Columbia University Press.

Arendt, Hannah. 1958. *The Human Condition*. Chicago: University of Chicago Press.

———. 1971. "Thinking and Moral Considerations." *Social Research* 38(3): 417–446.

Argyris, Chris, and Donald Schön. 1978. *Organizational Learning*. Reading, Mass.: Addison-Wesley.

Aristotle. 1962. *Nicomachean Ethics*. New York: Bobbs Merrill (Ostwald trans., 7th ed.).

Arnstein, Sherry. 1969. "A Ladder of Citizen Participation." *Journal of the American Institute of Planners* 35(4):216–224.

Austin, John. 1961. "A Plea for Excuses." In Austin, *Philosophical Papers*. London: Oxford University Press.

Bacow, Lawrence, and Michael Wheeler. 1984. *Environmental Dispute Resolution*. New York: Plenum Press.

Bailey, F. G. 1983. *The Tactical Uses of Passion*. Ithaca, N.Y.: Cornell University Press.

Barber, Benjamin. 1984. *Strong Democracy*. Berkeley: University of California Press.

Bar-On, Dan. 1990. *The Legacy of Silence*. Cambridge: Harvard University Press.

———. 1999. *The Indescribable and the Undiscussable: Reconstructing Human Discourse After Trauma*. Budapest: Central European University Press.

Barry, Brian, and Douglas Rae. 1975. "Political Evaluation." In F. Greenstein and N. Polsby, eds., *Handbook of Political Science,* Vol. 1. *Political Science: Scope and Theory*. Reading, Mass.: Addison-Wesley.

Batchelor, Peter, and David Lewis. 1986. *Urban Design in Action: The History, Theory and Development of the ATA Regional/Urban Design Assistance Teams Program*. Raleigh: North Carolina State University, School of Design.

Baum, Howell. 1983. *Planners and Public Expectations*. Cambridge, Mass.: Schenkman.

———. 1987. *The Invisible Bureaucracy: The Unconscious in Organizational Problem Solving*. New York: Oxford University Press.

———. 1990. *Organizational Membership*. Albany: State University of New York Press.

———. 1997a. *The Organization of Hope*. Albany: State University of New York Press.

———. 1997b. "Teaching Practice." *Journal of Planning Education and Research* 17(1):21–30.

Beauregard, Robert. 1989. "Between Modernity and Postmodernity: The Ambiguous Position of U.S. Planning." *Environment and Planning D: Society and Space* 7(4):381–395.

———. 1991. "Without a Net: Modernist Planning and the Postmodern Abyss." *Journal of Planning Education and Research* 10(3):189–194.

———. 1995. "Edge Critics." *Journal of Planning Education and Research* 14(3): 163–166.

Beiner, Ronald. 1983. *Political Judgment*. Chicago: University of Chicago Press.

Belenchy, Mary, et al. 1986. *Women's Ways of Knowing*. New York: Basic Books.

Benhabib, Seyla. 1985. *Critique, Norm, Utopia*. New York: Columbia University Press.

———. 1988. "Judgment and the Moral Foundations of Politics in Arendt's Thought." *Political Theory* 16(1):29–51.

———. 1989. "Liberal Dialogue Versus a Critical Theory of Discursive Legitimation." In N. Rosenblum, ed., *Liberalism and the Moral Life* (pp. 143–156). Cambridge: Harvard University Press.

———. 1990. "Hannah Arendt and the Redemptive Power of Narrative." *Social Research* 57(1):167–196.

———. 1992. *Situating the Self*. London: Routledge.

———. 1995. "Global Complexity, Moral Interdependence, and the Global Dialogical Community." In Martha Nussbaum and Jonathan Glover, eds., *Women, Culture, and Development: A Study of Human Capabilities* (pp. 235–255). Oxford: Clarendon Press.

———. 1996. *Democracy and Difference*. Princeton, N.J.: Princeton University Press.

———, and Fred Dellmayr, eds. 1990. *The Communicative Ethics Controversy*. Cambridge: MIT Press.

Bernstein, Richard. 1976. *The Restructuring of Social and Political Theory*. Philadelphia: University of Pennsylvania Press.

———. 1983. *Beyond Objectivism and Relativism*. Philadelphia: University of Pennsylvania Press.

Bessette, Joseph. 1994. *The Mild Voice of Reason: Deliberative Democracy and American National Government*. Chicago: University of Chicago Press.

Bingham, Gail. 1986. *Resolving Environmental Disputes: A Decade of Experience*. Washington, D.C.: Conservation Foundation.

Blackburn, J. Walton. 1988. "Environmental Mediation as an Alternative to Litigation." *Policy Studies Journal* 16(3):562–574.

Bok, Sissela. 1978. *Lying*. New York: Pantheon.

Bowles, Samuel, and Herbert Gintis. 1986. *Democracy and Capitalism: Property, Community, and the Contradictions of Modern Social Thought*. New York: Basic Books.

Bressi, Todd. 1994. "Planning the American Dream." In Peter Katz, *The New Urbanism*. New York: McGraw-Hill.

Brest, Paul. 1988. "Further Beyond the Republican Revival: Toward Radical Republicanism." *Yale Law Review* 97(8):1623–1631.

Brown, L. David, and Rajesh Tandon. 1983. "Ideology and Political Economy in Inquiry: Action Research and Participatory Research." *Journal of Applied Behavioral Science* 19(3):277–294.

Bryson, John, and Barbara Crosby. 1996. "Planning and the Design and Use of Forums, Arenas, and Courts." In Seymour Mandelbaum, Luigi Mazza, and Robert Burchell, eds. *Explorations in Planning Theory* (pp. 462–482). 1995: New Brunswick: Center for Urban Policy Research.

———. 1992. *Leadership for the Common Good: Tackling Public Problems When No One Is in Charge*. San Francisco: Jossey-Bass.

Burns, Robert P. 1989. "The Appropriateness of Mediation: A Case Study and Reflections on Fuller and Fiss." *Ohio State Journal on Dispute Resolution* 2(4): 129–155.

Burton, Lloyd. 1988. "Negotiating the Cleanup of Toxic Groundwater Contamination: Strategy and Legitimacy." *Natural Resources Journal* 28:105–143.

Calhoun, Craig. 1994. *Social Theory and the Politics of Identity*. Cambridge, Mass.: Blackwell.

Calthorpe, Peter. 1993. *The Next American Metropolis: Ecology, Community, and the American Dream*. Princeton, N.J.: Princeton Architectural Press.

Campbell, Scott, and Susan Fainstein, eds. 1996. *Readings in Planning Theory*. Cambridge: Blackwell.

Caro, Robert. 1998. "The City Shaper." *New Yorker,* Jan. 5, pp. 38–55.

Carpenter, Susan, and William Kennedy. 1988. *Managing Public Disputes*. San Francisco: Jossey-Bass.

Cavell, Stanley. 1968. *Must We Mean What We Say?* Cambridge: Cambridge University Press.

Cervero, Ronald M., and Arthur L. Wilson. 1994. *Planning Responsibly for Adult Education: A Guide to Negotiating Interests and Power*. San Francisco: Jossey-Bass.

Checkoway, Barry, ed. 1994. Symposium, "Paul Davidoff and Advocacy Planning in Retrospect." *Journal of the American Planning Association* 60:139–161.

Churchman, C. W. 1971. *The Design of Inquiring Systems*. New York: Basic Books.

Clavel, Pierre. 1985. *The Progressive City*. New Brunswick: Rutgers University Press.

Cobb, Michael D., and James H. Kuklinski. 1997. "Changing Minds: Political Arguments and Political Persuasion." *American Journal of Political Science* 41(1): 88–121.

Cohen, Joshua. 1993. "Moral Pluralism and Political Consensus." In D. Copp, J. Hampton, and J. E. Roemer, eds., *The Idea of Democracy*. Cambridge: Cambridge University Press.

Cohen, Michael D., James G. March, and Johan P. Olsen. 1988. "A Garbage Can Model of Organizational Choice." In James March, *Organizations and Decisions*. New York: Basil Blackwell.

Coleman, Jules. 1983. "Liberalism, Unfair Advantage, and the Volunteer Armed Forces." In Robert Fullinwider, eds., *Conscripts and Volunteers: Military Requirements, Social Justice, and the All Volunteer Force*. Totowa, N.J.: Rowman and Allenheld.

Coles, Robert. 1986. *The Moral Life of Children*. Boston: Atlantic Monthly Press.

———. 1989. *The Call of Stories*. Boston: Houghton Mifflin.

Cormick, Gerald. 1987. "The Myth, the Reality, and the Future of Environmental Mediation." In Robert Lake, ed., *Resolving Locational Conflict*. New Brunswick, N.J.: Center for Urban Policy Research.

Dallmayr, F., and T. McCarthy. 1977. *Understanding and Social Inquiry*. Notre Dame: University of Notre Dame Press.

Davy, Benjamin. 1997. *Essential Injustice*. New York: Springer.

Deetz, Stan. 1992. *Democracy in an Age of Corporate Colonization*. Albany: State University of New York Press.

Dewey, John. 1930. *Human Nature and Conduct*. New York: Henry Holt.

Dotson, Bruce, David Godschalk, and Jerome Kaufman. 1989. *The Planner as Dispute Resolver*. Washington, D.C.: National Institute for Dispute Resolution.

Drake, William. 1989. "Statewide Offices of Mediation." *Negotiation Journal* 5(4):359–364.

Dror, Yehezkel. 1992. "Israel Will Go in Traumas." *Jerusalem Journal of International Relations* 14(4).

Dryzek, John. 1990. *Discursive Democracy: Politics, Policy, and Political Science*. Cambridge: Cambridge University Press.

———. 1995. "The Informal Logic of Institutional Design." In Robert Goodin, ed., *The Theory of Institutional Design* (pp. 103–125). Cambridge: Cambridge University Press.

———. 1996. "From Irrationality to Autonomy: Two Sciences of Institutional Design." In S. Elkin and Karol Soltan, eds. *The Constitution of Good Societies*. Philadelphia: Pennsylvania State University Press.

Dudley, Mary Jo. 1996. "Uncovering the Invisible Workers: Using Participatory Video and Feminist Methodologies to Challenge Distorted Images of Colombian Domestic Workers." Master's thesis, Cornell University.

Dunn, William. 1981. *Introduction to Public Policy Analysis*. Englewood Cliffs, N.J.: Prentice Hall.

Dyckman, John. [ca. 1960]. "Introduction to Readings in the Theory of Planning: The State of Planning Theory in America." Unpublished manuscript.

Edelman, Murray. 1964. *The Symbolic Uses of Politics*. Urbana: University of Illinois Press.

———. 1988. *Constructing the Political Spectacle*. Chicago: University of Chicago Press.

Edwards, Harry. 1986. "Commentary: Alternative Dispute Resolution: Panacea or Anathema?" Harvard Law Review 99:668–684.

Elden, Max, and Morten Levin. 1990. "Co-generative Learning: Bringing Participation into Action Research." In W. F. Whyte, ed., *Participative Action Research*. Beverly Hills, Calif.: Sage.

Elkin, Stephen. 1987. *City and Regime in the American Republic*. Chicago: University of Chicago Press.

Elster, Jon. 1979. *Ulysses and the Sirens: Studies in Rationality and Irrationality*. New York: Cambridge University Press.

———. *Sour Grapes*. Cambridge: Cambridge University Press.

Esquith, Stephen L. 1994. *Intimacy and Spectacle: Liberal Theory as Political Education*. Ithaca, N.Y.: Cornell University Press.

Euben, Peter. 1990. *The Tragedy of Political Theory*. Princeton: Princeton University Press.

Faludi, Andreas. 1985. "The Return of Rationality." In M. Breheny and A. Hooper, eds., *Rationality in Planning*. London: Pion.

———. 1996. "Rationality, Critical Rationalism, and Planning Doctrine." In Seymour Mandelbaum, Luigi Mazza, and Robert Burchell, eds., *Explorations in Planning Theory*. New Brunswick, N.J.: Center for Urban Policy Research.

Fay, Brian. 1996. *Contemporary Philosophy of Social Science*. Oxford: Blackwell.

Feldman, Martha. 1989. *Order Without Design: Information Production and Policy Making*. Palo Alto: Stanford University Press.

Ferraro, Giovanni. 1992. "Irrationality in Planning." Typescript.

Fischer, Frank. 1980. *Politics, Values and Public Policy*. Boulder, Colo.: Westview Press.

———. 1985. "Critical Evaluation of Public Policy: A Methodological Case Study." In John Forester, ed., *Critical Theory and Public Life* (pp. 231–257). Cambridge: MIT Press.

———. 1995. *Evaluating Public Policy*. Chicago: Nelson-Hall.

Fischer, Frank, and John Forester, eds. 1987. *Confronting Values in Policy Analysis: The Politics of Criteria*. Los Angeles: Sage.

———, eds. 1993. *The Argumentative Turn in Policy Analysis and Planning*. Durham, N.C.: Duke University Press.

Fischler, Raphaël. 1995. "Planning Theory as Culture and Experience." Journal of Planning Education and Research 14(4):173–178.

Fisher, Roger, and William Ury. 1983. *Getting to Yes*. New York: Penguin.

Fishkin, James. 1991. *Democracy and Deliberation*. New Haven: Yale University Press.

Flyvbjerg, Bent. 1996. "The Dark Side of Planning: Rationality and Realrationalitat." In Seymour Mandelbaum, Luigi Mazza, and Robert Burchell, eds., *Explorations in Planning Theory*. New Brunswick, N.J.: Center for Urban Policy Research.

———. 1998. *Rationality and Power*. Chicago: University of Chicago Press.

Folberg, Jay, and Alison Taylor. 1984. *Mediation: A Comprehensive Guide to Resolving Conflicts Without Litigation*. San Francisco: Jossey-Bass.

Forester, John. 1980. "Listening: The Social Policy of Everyday Life (Critical Theory and Hermeneutics in Practice)." *Social Praxis* 7(3–4):219–232.

———.1982. "Public Policy and Respect." democracy 2(4):93–101.

———. 1985a. "Neither Handmaiden Nor Panacea: A Review of *Resolving Environmental Regulatory Disputes,* by Lawrence Susskind, Lawrence Bacow, and Michael Wheeler, eds. *Environmental Impact Assessment Review* 5(4):383–388.

———. 1985b. *Critical Theory and Public Life*. Cambridge: MIT Press.

———. 1987a. "Planning in the Face of Conflict." *Journal of the American Planning Association* 53:303–314.

———. 1987b. "Anticipating Implementation: Normative Practices in Planning and Policy Analysis." In Frank Fischer and John Forester, eds., *Confronting Values in Policy Analysis: The Politics of Criteria*. Los Angeles: Sage.

———. 1989. *Planning in the Face of Power*. Berkeley: University of California Press.

———. 1991a. "Mediated Negotiation in Urban Planning Practice: Practical Challenges and Theoretical Issues." Typescript.

———. 1991b. "Reply to Reviewers of *Planning in the Face of Power*." *International Planning Theory Newsletter*, Winter.

———. 1993a. "On Critical Ethnography: Fieldwork in a Habermasian Way." In M. Alvesson and H. Wilmott, eds., *Critical Management Studies*. Los Angeles: Sage.

———. 1993b. *Critical Theory, Public Policy, and Planning Practice*. Albany: State University of New York Press.

———. 1994a. "Dispute Resolution as a Strategy of Urban Planning: A European Exemplar, Profile of Rolf H. Jensen." Working Papers in Planning. Ithaca, N.Y.: Cornell University Department of City and Regional Planning.

———. 1994b. "Lawrence Susskind: Activist Mediation and Public Disputes." In Deborah M. Kolb et al., *When Talk Works: Profiles of Mediators*. San Francisco: Jossey-Bass.

———. 1995. "Community Land Use Planning via Consensus Building: A Profile of Bill Potapchuk." Working Papers in Planning. Ithaca, N.Y.: Cornell University, Department of City and Regional Planning.

———. 1997. *Learning from Practice: Democratic Deliberations and the Promise of Planning Practice*. College Park, Md.: Urban Studies and Planning Program.

———. 1998a. "Rationality, Dialogue, and Learning: What Community and Environmental Mediators Can Teach Us About the Practice of Civil Society." In Michael Douglass and John Friedmann, eds., *Cities for Citizens*. New York: Wiley.

———. 1998b. "Creating Public Value in Planning: The Three Abiding Problems of Negotiation, Participation, and Deliberation." *Urban Design International* 3(1):5–12.

———. 1999a. "Reflections on the Future Understanding of Planning Practice." *International Planning Studies*.

———. 1999b. "Dealing with Deep Value Differences: How Can Consensus Building Make a Difference?" In Lawrence Susskind, ed., *Handbook on Consensus-Building*. Beverly Hills Calif.: Sage.

———. Forthcoming. "An Instructive Case Hampered by Theoretical Puzzles: Critical Comments on Bent Flyvbjerg's *Rationality and Power*." *International Planning Theory*.

Forester, John, and L. Chu, eds. 1990. "Profiles of Planners." Typescript. Working Papers, 106 W. Sibley Hall, Cornell University, Ithaca, New York. 14853, 234 pages.

Forester, John, Raphaël Fischler, and Deborah Shmueli, eds. 1997. *Profiles of Community Builders: Israeli Planners and Designers*. Haifa, Israel: Klutznick Center, The Technion.

Forester, John, and B. Kreiswirth, eds. 1993a. "Profiles of Planners in Housing and Community Development." Typescript. Profile/Working Papers, 106 W. Sibley Hall, Cornell University, Ithaca, New York. 14853, 209 pages.

———. 1993b. "Profiles of Planners in Land Use, Transportation and Environmental Planning." Typescript. Profile/Working Papers, 106 W. Sibley Hall, Cornell University, Ithaca, New York. 14853, 174 pages.

———. 1993c. "Profiles of Planners in Historic Preservation Planning." Typescript. Profile/Working Papers, 106 W. Sibley Hall, Cornell University, Ithaca, New York. 14853, 166 pages.

———. 1993d. "Profiles of Women in Planning." Typescript. Profile/Working Papers, 106 W. Sibley Hall, Cornell University, Ithaca, New York. 14853, 157 pages.

Forester, John, J. Pitt, and J. Welsh, eds. 1993. "Profiles of Participatory Action Researchers." Typescript. Profile/Working Papers, 106 W. Sibley Hall, Cornell University, Ithaca, New York. 14853. 191 pages.

Forester, John, and D. Stitzel. 1989a. "Beyond Neutrality: The Possibilities of Activist Mediation in Public Sector Conflicts." *Negotiation Journal* July:251–264.

———. 1989b. "Westville: Mediation Strategies in Community Planning" and "Teaching Notes for Westville." Available from the Harvard Program on Negotiation. Cambridge, Mass.

Forester, John, and I. Weiser, eds. 1996. "Making Mediation Work: Profiles of Environmental and Community Mediators." Typescript. Profile/Working Papers, 106 W. Sibley Hall, Cornell University, Ithaca, New York. 14853. 176 pages.

Foucault, Michel. 1980. *Power/Knowledge: Selected Interviews and Other Writings, 1972–1977*. Ed. Colin Gordon. New York: Pantheon.

Freire, Paulo. 1970. *Pedagogy of the Oppressed*. New York: Seabury Press.

Fried, Charles. 1976. "The Lawyer as Friend: The Moral Foundations of the Lawyer-Client Relationship." *Yale Law Journal* 85:1060.

Friedlander, Saul. 1992. "Trauma, Transference, and Working Through." *History and Memory* 4:39–55.

Friedman, Marilyn. 1989. "Feminism and Modern Friendship: Dislocating the Community." *Ethics* 99(2):275–290.

Friedmann, John. 1987. *Planning in the Public Domain*. Princeton, N.J.: Princeton University Press.

Frug, Gerald. 1988. "Argument as Character." *Stanford Law Review* 40:869–927.

Garcia, Angela. 1991. "Dispute Resolution Without Disputing: How the Interactional Organization of Mediation Hearings Minimizes Argument." *American Sociological Review* 56:818–835.

Gaventa, John. 1980. *Power and Powerlessness*. Urbana: University of Illinois Press.

———. 1991. "Toward a Knowledge Democracy: Viewpoints on Participatory Research in North America." In Orlando Fals-Borda and Mohammed Rahman, eds., *Action and Knowledge: Breaking the Monopoly with Participatory Action Research*. New York, Apex Press.

———, and Billy Horton. 1981. "A Citizen's Research Project in Appalachia, USA." *Convergence* 14:3.

George, R. Varkki. 1994. "Formulating the Right Planning Problem." *Journal of Planning Literature* 8(3):240–259.

Giddens, Anthony. 1984. *The Constitution of Society*. Berkeley: University of California Press.

Gilligan, Carol. 1982. *In a Different Voice: Psychological Theory and Women's Development*. Cambridge: Harvard University Press.

Glass, James. 1989. *Private Terror, Public Life*. Ithaca, N.Y.: Cornell University Press.

Goffman, Erving. 1967. *Interaction Ritual*. Garden City, N.Y.: Anchor Books.

Goldschmidt, Gabriela. 1991. "The Dialectics of Sketching." *Creativity Research Journal* 4(2):123–143.

Goodman, Nelson. 1978. *Ways of Worldmaking*. Cambridge, Mass.: Hackett.

Goodman, Paul. 1951. *Utopian Essays and Practical Proposals*. New York: Vintage.

Grad, Frank. 1989. "Alternative Dispute Resolution in Environmental Law." *Columbia Journal of Environmental Law* 14:157–185.

Granovetter, Marc. 1985. "Economic Action and Social Structure: The Problem of Embeddedness." *American Journal of Sociology* 91(3):481–510.

Grant, Jill. 1994. *The Drama of Democracy: Contention and Dispute in Community Planning*. Toronto: University of Toronto Press.

Greenwood, Davydd. 1991. "Collective Reflective Practice Through Participatory Action Research: A Case Study from the Fagor Cooperatives of Mondragon." In D. Schön, ed., *The Reflective Turn*. New York: Teacher's College Press.

———, and Morten Levin. 1999. *Introduction to Action Research: Social Research for Social Change*. Thousand Oaks, Calif.: Sage.

Gusfield, Joseph. 1981. *The Culture of Public Problems*. Chicago: University of Chicago Press.

———, ed. 1989. *Kenneth Burke: On Symbols and Society*. Chicago: University of Chicago Press.

Gustavsen, Bjorn. 1992. *Dialogue and Development: Theory of Communication, Action Research, and the Development of Working Life*. Stockholm: Center for Swedish Working Life.

———, and Stephen Toulmin. 1996. *Beyond Theory: Changing Organizations Through Participation*. Amsterdam: John Benjamins Publishing Co.

Gutmann, Amy. 1985. "Communitarian Critics of Liberalism." *Philosophy and Public Affairs* 14(3):308–322.

———. 1993a. "The Challenge of Multiculturalism in Political Ethics." *Philosophy and Public Affairs* 22(3):171–206.

———. 1993b. "The Disharmony of Democracy." In J. W. Chapman and I. Shapiro, eds., *Democratic Community*. New York: New York University Press.

Gutmann, Amy, and Charles Taylor. 1992. *Multiculturalism and "The Politics of Recognition."* Princeton, N.J.: Princeton University Press.

Gutmann, Amy, and Dennis Thompson. 1996. *Democracy and Disagreement: Why Moral Conflict Cannot be Avoided in Politics, and What Should Be Done About It*. Cambridge: Harvard University Press.

Guttman, Nurit. 1996. "Values and Justifications in Health Communication Interventions: An Analytic Framework." *Journal of Health Communication* 1:365–396.

Habermas, Jürgen. 1975. *Legitimation Crisis*. Boston: Beacon Press.

———. 1979. *Communication and the Evolution of Society*. Boston: Beacon Press.

———. 1984. *The Theory of Communicative Action*. Boston: Beacon Press.

———. 1990. *Moral Consciousness and Communicative Action*. Cambridge: MIT Press.

———. 1996. *Between Facts and Norms*. Cambridge: MIT Press.

Harrington, C. 1985. *Shadow Justice: The Ideology and Institutionalization of Alternatives to Court*. Westport, Conn. Greenwood.

Harter, Philip. 1982. "Negotiating Regulations: A Cure for Malaise." *Georgetown Law Journal* 71(1):11–118.

———. 1989. "A General Overview of Negotiated Rulemaking and Other Forms of Administrative Dispute Resolution." Typescript. First Annual Review of the Administrative Process. Washington, D.C.: Federal Bar Association and Washington College of Law, June 14.

Hartman, Chester. 1975. "The Advocate Planner: From Hired Gun to Political Partisan." In R. A. Cloward and F. F. Piven, eds., *The Politics of Turmoil*. New York: Vintage.

———. 1978. "Social Planning and the Political Planner." In Robert Burchell and George Sternlieb, eds., *Planning Theory in the 1980's*. New Brunswick, N.J.: Center for Urban Policy Research.

———. 1981. "The Limits of Consensus Building." In J. DeNeufville, *The Land Use Policy Debate in the United States* (pp. 205–207). New York: Plenum Press.

Hayden, Dolores, 1984. *Redesigning the American Dream: The Future of Housing, Work, and Family Life*. New York: Norton.

Healey, Patsy. 1992. "A Day's Work: Knowledge and Action in Communicative Practice." *Journal of the American Planning Association* 58:9–20.

———. 1993a. "The Communicative Work of Development Plans." *Environment and Planning B: Planning and Design* 20:83–104.

———. 1993b. "Planning Through Debate: The Communicative Turn in Planning Theory." In F. Fischer and J. Forester, eds., *The Argumentative Turn in Policy Analysis and Planning* (pp. 233–253). Durham, N.C.: Duke University Press.

———. 1996. "The Communicative Turn in Planning Theory and Its Implications for Spatial Strategy Formation." *Environment and Planning B: Planning and Design* 23:217–234.

———. 1997. *Collaborative Planning: Making Frameworks in Fragmented Societies*. London: Macmillan.

Healey, Patsy, and J. Hillier. 1996. "Communicative Micropolitics." *International Planning Studies* 1:2.

Healey, Patsy, P. McNamara, M. Elson, and J. Doak. 1988. *Land Use Planning and the Mediation of Urban Change*. Cambridge: Cambridge University Press.

Heidegger, Martin. 1962. *Being and Time*. New York: Harper & Row.

Hendler, Sue. 1995. *Planning Ethics*. New Brunswick, N.J.: Center for Urban Policy Research.

Heritage, John. 1984. *Garfinkel and Ethnomethodology*. Cambridge: Polity Press.

Herman, Judith. 1992. *Trauma and Recovery*. New York: Basic Books.

Hirschman, Albert. 1986. *Rival Views of Market Society and Other Recent Essays*. New York: Viking.

Hoch, Charles. 1988. "Conflict at Large: A National Survey of Planners and Political Conflict." *Journal of Planning Education and Research* 8(1):25–34.

———. 1994. *What Planners Do?* Chicago: APA Planners Press.

Honneth, Axel, and Charles Wright, eds. 1995. *The Fragmented World of the Social*. Albany: State University of New York Press.

Horton, Myles, and Paulo Freire. 1990. *We Make the Road By Walking*. Edited by Brenda Bell, John Gaventa, and John Peters. Philadelphia: Temple University Press.

Howe, Elizabeth. 1994. *Acting on Ethics in City Planning*. New Brunswick, N.J.: Center for Urban Policy Research.

Hummel, R. 1991. "Stories Managers Tell: Why They Are as Valid as Science." *Public Administration Review* 51(1):31–41.

Innes, Judith. 1995a. "Planning Is Institutional Design." *Journal of Planning Education and Research* 14:2.

———. 1995b. "Planning Theory's Emerging Paradigm." *Journal of Planning Education and Research* 14(3):183–189.

———. 1996. "Planning Through Consensus Building: A New View of the Comprehensive Ideal." *Journal of the American Planning Association* 62(4):460–472.

Innes, Judith, Judith Gruber, Michael Neuman, and Robert Thompson. 1994. *Coordinating Growth and Environmental Management Through Consensus Building*. Berkeley: Policy Research Program Report of the California Policy Seminar.

Irwin, Terrence. 1989. *Classical Thought*. Oxford: Oxford University Press.

Isaacs, William. 1993. "Taking Flight: Dialogue, Collective Thinking, and Organizational Learning." Report of the Center for Organizational Learning's Dialogue Project, MIT.

Jaffe, Sanford. 1984. "The Courts and Dispute Resolution." *Resolve* Winter: 2–3.

Jennings, B. 1990. "Ethics and Ethnography in Neonatal Intensive Care." In George Weisz, ed., *Social Science Perspectives on Medical Ethics*. London: Kluwer.

Johnson, Jeffrey Paul. 1986. "Negotiating Environmental and Development Disputes." *Journal of Planning Literature* 1(4):509–521.

Johnson, Ralph, and J. Anthony Blair. 1985. "Informal Logic: The Past Five Years, 1978–1983." *American Philosophical Quarterly* 22(3).

Kahneman, D. J., Knetsch, and R. Thaler. 1986. "Fairness as a Constraint on Profit Seeking: Entitlements in the Market." *American Economic Review* 76: 728–741.

Kahneman, D., and Tversky, A. 1979. "Prospect Theory: An Analysis of Decision Under Risk." *Econometrica* 47:263–291.

Katz, Peter. 1994. *The New Urbanism*. New York: McGraw-Hill.

Kelman, Herbert C. 1992. "Informal Mediation by the Scholar/Practitioner." In J. Bercovitch and J. Rubin, eds., *Mediation in International Relations* (pp. 64–96). New York: St. Martin's Press.

Kemmis, Daniel. 1990. *Community and the Politics of Place*. Norman: University of Oklahoma Press.

Kemmis, Stephen, and Wilford Carr. 1986. *Becoming Critical: Education, Knowledge, and Action Research*. New York: Taylor and Francis.

Kertzer, David. 1989. *Ritual, Politics, and Power*. New Haven: Yale University Press.

Kirp, David, et al. 1989. *Learning by Heart: AIDS and Schoolchildren in America's Communities*. New Brunswick, N.J.: Rutgers University Press.

Kittay, Eva, and Diana Meyers. 1987. *Women and Moral Theory*. Totowa, N.J.: Rowman and Littlefield.

Kochman, Thomas. 1981. *Black and White Styles in Conflict*. Chicago: University of Chicago Press.

Kolb, Deborah, ed. 1994. *When Talk Works*. San Francisco: Jossey-Bass.

———, and Jeffrey Rubin. 1989. "Mediation Through a Disciplinary Kaleidoscope: A Summary of Empirical Research." *Dispute Resolution Forum (NIDR)* October:3–8.

Kraft, Michael E. 1996. "Democratic Dialogue and Acceptable Risks: The Politics of High-Level Nuclear Waste Disposal in the United States." In Don Munton, ed., *Hazardous Waste Siting and Democratic Choice* (pp. 108–141). Washington, D.C.: Georgetown University Press.

Kraybill, Ron. 1994. "Neutralising History." *Track Two* September:39–40.

Kressel, Kenneth, et al. 1989. *Mediation Research: The Process and Effectiveness of Third Party Intervention*. San Francisco: Jossey-Bass.

Krieger, M. 1981. *Advice and Planning*. Philadelphia: Temple University Press.

Kronman, A. 1987. "Living in the Law." *Chicago Law Review* 54:835.

———. 1986–1987. "Practical Wisdom and Professional Character." *Social Philosophy and Policy* 4(1):203–234.

Krumholz, Norman. 1982. "A Retrospective View of Equity Planning: Cleveland 1969–1979." *Journal of the American Planning Association* 48:163–174.

Krumholz, Norman, and Pierre Clavel. 1994. *Reinventing Cities*. Philadelphia: Temple University Press.

Krumholz, Norman, and J. Forester. 1990. *Making Equity Planning Work*. Philadelphia: Temple University Press.

LaCapra, Dominick. 1983. "Bakhtin, Marxism, and the Carnivalesque." In *Rethinking Intellectual History*. Ithaca, N.Y.: Cornell University Press.

———. 1994. *Representing the Holocaust: History, Theory, and Trauma*. Ithaca, N.Y.: Cornell University Press.

Laue, James, and Gerald Cormick. 1978. "The Ethics of Intervention in Community Disputes." In G. Bermant, H. Kelman, and D. Warwick, eds., *The Ethics of Social Intervention* (pp. 205–232). Washington, D.C.: Halstead Press.

Lax, David, and James Sebenius. 1987. *The Manager as Negotiator*. New York: Free Press.

Levin, Morten. 1993. "Creating Networks for Rural Economic Development in Norway." *Human Relations* 46(2):193–218.

Lindblom, Charles. 1959. "The Science of Muddling Through." *Public Administration Review* 19:79–88.

———. 1990. *Inquiry and Change*. New Haven, Conn.: Yale University Press.

Lowry, Kem, P. Adler, and N. Milner. 1997. "Participating the Public: Group Process, Politics, and Planning." *Journal of Planning Education and Research* 16: 177–187.

Luban, David. 1983. "Explaining Dark Times: Hannah Arendt's Theory of Theory." *Social Research* 50(1):215–248.

———. 1985. "Bargaining and Compromise: Recent Work on Negotiation and Informal Justice." *Philosophy and Public Affairs* 14(4):397–416.

Lukes, Steven. 1974. *Power: A Radical View*. New York: Macmillan.

———. 1975. "Political Ritual and Social Integration." *Sociology* 9(2):289–308.

Lyons, David. 1984. *Ethics and the Rule of Law*. Cambridge: Cambridge University Press.

MacIntyre, Alasdair. 1981. *After Virtue*. Notre Dame, Ind.: University of Notre Dame Press.

Majone, G. 1989. *Evidence, Argument, and Persuasion in the Policy Process*. New Haven, Conn.: Yale University Press.

Mandelbaum, Seymour. 1991. "Telling Stories." *Journal of Planning Education and Research* 10:209–214.

———, Luigi Mazza, and Robert Burchell, eds., 1996. *Explorations in Planning Theory*. New Brunswick, N.J.: Center for Urban Policy Research.

Manin, B. 1987. "On Legitimacy and Political Deliberation." *Political Theory* August: 338–368.

Mansbridge, Jane. 1988. *Beyond Self Interest*. Chicago: University of Chicago Press.

———. 1992. "A Deliberative Theory of Interest Representation." In Mark Petracca, ed., *The Politics of Interests*. Boulder, Colo.: Westview.

March, James. 1988. *Organizations and Decisions*. New York: Blackwell.

March, James, and Johann Olsen. 1976. *Ambiguity and Choice in Organizations*. Universitetsforlaget, Oslo.

Margalit, Avishai. 1993. "Prophets with Honor." *New York Review of Books*, November 4, pp. 66–71.

Marris, Peter. 1975. *Loss and Change*. New York: Anchor.

———. 1982. *Community Planning and Conceptions of Change*. London: Routledge.

———. 1990. "Witnesses, Engineers, or Story-Tellers? The Influence of Social Research on Social Policy." In Herbert Gans, ed., *Sociology in America*. Los Angeles: Sage.

———. 1996. *The Politics of Uncertainty*. London: Routledge.

Mathews, David. 1994. *Politics for People: Finding a Responsible Public Voice*. Urbana: University of Illinois Press.

McClendon, Bruce W., and A. Catanese, eds. 1996. *Planners on Planning*. San Francisco: Jossey-Bass.

McCloskey, Donald. 1985. *The Rhetoric of Economics*. Madison: University of Wisconsin Press.

Meer, Elizabeth. 1989. "Environmental Mediation: Toward a New Model." Master's thesis, Cornell University.

Menkel-Meadows, Carrie. 1995. "The Many Ways of Mediation: The Transformation of Traditions, Ideologies, Paradigms, and Practices." *Negotiation Journal* July:217–242.

Michelman, Frank. 1988. "Law's Republic." *Yale Law Review* 97(8):1493–1537.

Miller, David. 1992. "Deliberative Democracy and Social Choice." *Political Studies* 40 (special issue):54–67.

Miller, William Ian. 1993. *Humiliation*. Ithaca, N.Y.: Cornell University Press.

Mills, C. Wright. 1959. *The Sociological Imagination*. New York: Oxford University Press.

Minard, Richard, Ken Jones, and Christopher Patterson. 1993. "State Comparative Risk Projects: A Force for Change." Northeast Center for Comparative Risk, South Royalton: Vermont Law School.

Mischler, Elliot. 1985. *The Discourse of Medicine: The Dialectics of Medical Interviews*. Norwood, N.J.: Ablex.

Montville, Joseph. 1991. "Psychoanalytic Enlightenment and the Greening of Diplomacy." In V. Volkan, J. Montville, and D. Julius, eds., *The Psychodynamics of International Relationships,* Vol. 2: *Unofficial Diplomacy at Work*. Lexington, Mass.: Lexington Books.

Moore, Christopher. 1986. *The Mediation Process*. San Francisco: Jossey-Bass.

Moore, Mark. 1995. *Creating Public Value*. Cambridge: Harvard University Press.

Moore, Sally, and Barbara Myerhoff. 1977. *Secular Ritual*. Atlantic Highlands, N.J.: Humanities Press.

Mouffe, Chantal. 1993. *The Return of the Political*. London: Verso.

Mumby, Dennis K. 1988. *Communication and Power in Organizations: Discourse, Ideology, and Domination*. Norwood, N.J.: Ablex.

Murdoch, Iris. 1970. *The Sovereignty of Good*. London: Ark.

Myerhoff, Barbara. 1982. "Life History Among the Elderly: Performance, Visibility, and Re-Membering." In J. Ruby, ed., *A Crack in the Mirror: Reflexive Perspectives in Anthropology*. Philadelphia: University of Pennsylvania Press.

———. 1988. "Surviving Stories: Reflections on *Number Our Days*." In Jack Kugelmass, ed., *Between Two Worlds*. Ithaca, N.Y.: Cornell University Press.

Needleman, Carolyn, and Needleman, Martin. 1974. *Guerrillas in the Bureaucracy*. New York: Wiley.

Nelessen, Anton Clarence. 1994. *Visions for a New American Dream: Process, Principles, and an Ordinance to Plan and Design Small Communities*. Chicago: Planners Press, American Planning Association.

Neustadt, Richard, and Ernst May. 1986. *Thinking in Time*. New York: Free Press.

Nussbaum, Martha. 1986. *The Fragility of Goodness*. Cambridge: Cambridge University Press.

———. 1990a. *Love's Knowledge*. New York: Oxford University Press.

———. 1990b. "Aristotelian Social Democracy." G. Mara and H. Richardson, eds. In *Liberalism and the Good*. New York: Routledge.

———, and Amartya Sen, eds. 1993. *The Quality of Life*. Oxford: Oxford University Press.

O'Connor, James. 1973. *The Fiscal Crisis of the State*. New York: St. Martin's Press.

O'Neill, John. 1972. *Sociology as a Skin Trade*. New York: Harper & Row.

———. 1974. *Making Sense Together*. New York: Harper & Row.

———. 1989. *The Communicative Body*. Evanston, Ill.: Northwestern University Press.

Paris, David, and James Reynolds. 1983. *The Logic of Policy Inquiry*. New York: Longman.

Park, Peter, Mary Brydon-Miller, Budd Hall, and Ted Jackson, eds. 1993. *Voices of Change*. Westport, Conn.: Bergin and Garvey.

Peattie, Lisa. 1987. *Planning: Rethinking Ciudad Guayana*. Ann Arbor: University of Michigan Press.

Pitkin, Hanna. 1972. *Wittgenstein and Justice*. Berkeley: University of California Press.

———. 1981. "Justice: On Relating Private and Public." *Political Theory* 9(3): 327–352.

———. 1984. *Fortune Is a Woman: Gender and Politics in the Thought of Niccolo Machiavelli*. Berkeley: University of California Press.

Pitt, Jessica. 1993. "Beyond Rules and Theory: Linda Stout's Community Organizing Practice." Typescript. Paper for CRP642, Cornell University, Ithaca, N.Y.

Pops, Gerald, and Max Stephenson. 1988. "Public Administrators and Conflict Resolution: Problems and Prospects." *Policy Studies Journal* 16(3):615–626.

Putnam, Hilary. 1990. *Realism with a Human Face*. Cambridge: Harvard University Press.

———. 1987. *The Many Faces of Realism*. LaSalle, Ill.: Open Court.

Putnam, Robert. 1993. *Making Democracy Work*. Princeton: Princeton University Press.

Raiffa, Howard. 1985. *The Art and Science of Negotiation*. Cambridge: Harvard University Press.

———. 1985. "Post-Settlement Settlements." *Negotiation Journal* 1(1):9–12.

Raymond, Janice. 1986. *A Passion for Friends*. Boston: Beacon Press.

Reardon, Ken, John Welsh, Brian Kreiswirth, and John Forester. 1993. "Participatory Action Research from the Inside: A Profile of Ken Reardon's Community Development Practice in East St. Louis." *American Sociologist* 24(1):69–91.

———. 1993. "Community Planning in East St. Louis, Illinois: A Participatory Action Research Approach." Paper presented at the 1993 American Sociological Association Meeting, Miami Beach, Fla.

Reich, Robert. 1988. "Policymaking in a Democracy." In R. Reich, ed., *The Power of Public Ideas*. Cambridge: Ballinger.

Riccucci, Norma. 1995. *Unsung Heroes*. Washington, D.C.: Georgetown University Press.

Richards, Alun. 1996. "Using Co-management to Build Community Support for Waste Facilities." In Don Munton, ed., *Hazardous Waste Siting and Democratic Choice* (pp. 321–337). Washington, D.C.: Georgetown University Press.

Richardson, Henry. 1990. "Specifying Norms as a Way to Resolve Concrete Ethical Problems." *Philosophy and Public Affairs* 19:279–310.

Riddick, W. 1971. *Charrette Processes: A Tool in Urban Planning*. York, Pa.: George Shumway Publisher.

Rivkin, Malcolm. 1977. *Negotiated Development: A Breakthrough in Environmental Controversies*. Washington, D.C.: Conservation Foundation.

Roe, Emery. 1994. *Narrative Policy Analysis*. Durham, N.C.: Duke University Press.

Rorty, Amelie. 1988. *Mind in Action: Essays in the Philosophy of Mind*. Boston: Beacon Press.

Rorty, Richard. 1982. "Method, Social Science, and Social Hope." In *Consequences of Pragmatism* (pp. 191–210). Minneapolis: University of Minnesota Press.

Rosen, Michael. 1985. "Breakfast at Spiro's: Dramaturgy and Dominance." *Journal of Management* 11(2):31–48.

Rosen, Robert E. 1989. "Participation, Due Process, and Responsive Administration: Handler's *The Conditions of Discretion*." *Law and Social Inquiry* 14(2): 323–359.

Sager, Tore. 1994a. *Communicative Planning Theory*. Aldershot: Avebury Press.

———. 1994b. "Planning for the Paretian Liberal: A Democratic Dilemma in Social Choice Theory and Planning." Typescript.

Sandel, Michael. 1982. *Liberalism and the Limits of Justice*. Cambridge: Cambridge University Press.

Sandercock, Leonie. 1995. "Voices from the Borderlands: A Meditation on a Metaphor." *Journal of Planning Education and Research* 14(2):77–88.

———, and Ann Forsyth. 1992. "A Gender Agenda: New Directions for Planning Theory." *Journal of the American Planning Association* 58(1):49–59.

Sanders, Lynn. 1997. "Against Deliberation." *Political Theory* 25(3):347–376.

Santner, Eric L. 1992. "History Beyond the Pleasure Principle: Some Thoughts on the Representation of Trauma." In Saul Friedlander, ed., *Probing the Limits of Representation: Nazism and the "Final Solution"* (pp. 143–154). Cambridge: Harvard University Press.

Sarat, Austin. 1986. "Law and Strategy in the Divorce Lawyer's Office." *Law and Society Review* 20(1):93–134.

Schattschneider, E. E. 1960. *The Semi-Sovereign People: A Realist's View of Democracy in America*. New York: Holt, Rinehart, and Winston.

Schelling, Thomas. 1984. *Choice and Consequence*. Cambridge, Mass.: Harvard University Press.

Schneekloth, Lynda H., and Robert G. Shibley. 1995. *Placemaking: The Art and Practice of Building Communities*. New York: Wiley.

Schön, Donald. 1983. *The Reflective Practitioner: How Professionals Think in Action*. New York: Basic Books.

———. 1991. *The Reflective Turn: Case Studies in and on Educational Practice*. New York: Teacher's College Press.

———. 1992. "The Theory of Inquiry: Dewey's Legacy to Education." *Curriculum Inquiry* 22(2):119–139.

Seeley, John. 1963. "Social Science: Some Probative Problems." In A. Vidich and M. Stein, eds., *Sociology on Trial*. Englewood Cliffs, N.J.: Prentice-Hall.

Shoshkes, Ellen. 1989. *The Design Process: Case Studies in Project Development*. New York: Whitney Library of Design.

Silbey, Susan, and Austin Sarat. 1989. "Dispute Processing in Law and Legal Scholarship: From Institutional Critique to the Reconstruction of the Juridical Subject." *Denver University Law Review* 66(3):437–498.

Silbey, Susan, and Sally Merry. 1986. "Mediator Settlement Strategies." *Law and Policy* 8(1):7–32.

Smith, Bruce. 1985. *Politics and Remembrance*. Princeton, N.J.: Princeton University Press.

Smith, William P. 1985. "Effectiveness of the Biased Mediator." *Negotiation Journal* October:363–372.

Stacey, Judith. 1988. "Can There Be a Feminist Ethnography?" *Women's Studies International Forum* 11(1):21–27.

Storper, Michael, and Andrew Sayer. 1997. "Ethics Unbound: For a Normative Turn in Social Theory." *Environment and Planning D: Society and Space* 15:1–17.

Stulberg, Joseph. 1981. "The Theory and Practice of Mediation: A Reply to Professor Susskind." *Vermont Law Review* 6(1):85–117.

Sullivan, William. 1986. *Reconstructing Public Philosophy*. Berkeley: University of California Press.

Sunstein, Cass. 1988. "Beyond the Republican Revival." *Yale Law Review* 97(8): 1539–1590.

Susskind, Lawrence. 1981. "Environmental Mediation and the Accountability Problem." *Vermont Law Review* 6:1–47.

———. 1986a. "Evaluating Dispute Resolution Experiments." *Negotiation Journal* April:135–139.

———. 1986b. "NIDR's State Office of Mediation Experiment." *Negotiation Journal* October:323–327.

Susskind, Lawrence, Lawrence Bacow, and Michael Wheeler. 1983. *Resolving Environmental Regulatory Disputes*. Cambridge, Mass.: Schenkman.

Susskind, Lawrence, and J. Cruickshank. 1987. *Breaking the Impasse*. New York: Basic Books.

Susskind, Lawrence, and Patrick Field. 1996. *Dealing with an Angry Public*. New York: Free Press.

Susskind, Lawrence, and Denise Madigan. 1984. "New Approaches to Resolving Disputes in the Public Sector." *Justice System Journal* 9(2):179–203.

Susskind, Lawrence, Sara McKearnan, and Jennifer Thomas-Larmer, eds. 1999. *The Consumer Building Handbook*, Los Angeles: Sage.

Susskind, Lawrence, and Connie Ozawa. 1984. "Mediated Negotiation in the Public Sector: The Planner as Mediator." *Journal of Planning Education and Research* 4(1):5–15.

———. 1983. "Mediated Negotiation in the Public Sector: Mediator Accountability and the Public Interest Problem." *American Behavioral Scientist* 27(2): 255–279.

Szanton, Peter. 1981. *Not Well Advised*. New York: Russell Sage.

Taylor, Charles. 1989a. "Cross-Purposes: The Liberal-Communitarian Debate." In Nancy Rosenblum, ed., *Liberalism and the Moral Life*. Cambridge: Harvard University Press.

———. 1989b. *Sources of the Self*. Cambridge: Harvard University Press.

Taylor, Nigel. 1998. "Mistaken Interests and the Discourse Model of Planning." *Journal of the American Planning Association* 64(1):64–75.

Teitz, Michael B. 1996. "American Planning in the 1990s: Evolution, Debate, and Challenge." *Urban Studies* 33(4–5):649–671.

Tett, Alison, and Jeanne M. Wolfe. 1991. "Discourse Analysis and City Plans." *Journal of Planning Education and Research* 10(3):195–200.

Thompson, John. 1984. *Studies in the Theory of Ideology*. Berkeley: University of California Press.

Throgmorton, James. 1996. *Planning as Persuasive Storytelling: The Rhetorical Construction of Chicago's Electric Future*. Chicago: University of Chicago Press.

Tribe, Laurence. 1973. "Technology Assessment and the Fourth Discontinuity." *Southern California Law Review* 46:617–660.

Tronto, Joan. Summer 1987. "Beyond Gender Difference to a Theory of Care." *Signs: Journal of Women in Culture and Society* 12(4):644–663.

Turner, Victor. 1969. *The Ritual Process: Structure and Anti-Structure*. London: RKP.

———. 1974. *Dramas, Fields and Metaphors: Symbolic Action in Human Society*. Ithaca, N.Y.: Cornell University Press.

Ury, William, Jeanne Brett, and Stephen Goldberg. 1988. *Getting Disputes Resolved*. San Francisco: Jossey-Bass.

Van Maanen, John. 1988. *Tales of the Field: On Writing Ethnography*. Chicago: University of Chicago Press.

Ventriss, Curtis. 1991. "Reconstructing Government Ethics: A Public Philosophy of Civic Virtue." In J. Bowman, ed., *Ethical Frontiers in Public Management* (pp. 114–134). San Francisco: Jossey-Bass.

Vickers, S. G. 1995. *The Art of Judgment*. Los Angeles: Sage.

Viggiani, Frances. 1991. Democratic Hierarchies in the Workplace. Ph.D. dissertation, Cornell University.

Waldron, Jeremy. 1988. "When Justice Replaces Affection: The Need for Rights." *Harvard Journal of Law and Public Policy* 11(3):625–647.

Wallace, James. 1988. *Moral Relevance and Moral Conflict*. Ithaca, N.Y.: Cornell University Press.

Wardhaugh, Ronald. 1985. *How Conversation Works*. New York: Basil Blackwell.

Warren, Mark. 1992. "Democratic Theory and Self-Transformation." *American Political Science Review* 86:8–23.

Webber, M., and H. Rittel. 1971. "Dilemmas in a General Theory of Planning." *Policy Sciences* 4:155–169.

White, James Boyd. 1985. "Law as Rhetoric, Rhetoric as Law: The Arts of Cultural and Communal Life." *University of Chicago Law Review* 52(3):684–702.

Wiggins, David. 1978. "Deliberation and Practical Reason." In J. Raz, ed., *Practical Reasoning*. Oxford: Oxford University Press.

Williams, Bruce A., and Albert R. Matheny. 1995. *Democracy, Dialogue, and Environmental Disputes: The Contested Languages of Social Regulation*. New Haven: Yale University Press.

Willis, Paul. 1977. *Learning to Labor*. New York: Columbia University Press.

Winner, Langdon. 1986. *The Whale and the Reactor: A Search for Limits in an Age of High Technology*. Chicago: University of Chicago Press, 1986.

Wittgenstein, Ludwig. 1967. *Philosophical Investigations*. Oxford: Blackwell.

Yankelovich, Daniel. 1991. *Coming to Public Judgment*. Syracuse, N.Y.: Syracuse University Press.

Yiftachel, Oren. 1995. "Planning as Control: Policy and Resistance in a Deeply Divided Society." *Progress in Planning* 44:116–187.

Young, Iris Marion. 1990. *Justice and the Politics of Difference*. Princeton, N.J.: Princeton University Press.

Index